Sustainable Nanomaterials for the Construction Industry

Sustainable Nanomaterials for the Construction Industry examines applications of sustainable nanomaterials used in the building construction sector. The chapters focus on sustainable construction materials using nanotechnology such as pigments, modified cement, polymer, glass, phase-change materials and air purification.

- Highlights nanotechnology applications in smart buildings
- Reviews nano-enhanced glass and phase-change materials for energy saving and energy storage
- Discusses nanomaterials used in air purification applications as well as sustainable pigments
- Covers latest developments in polymers, glasses, coatings, paints and insulating materials

Aimed at materials and construction engineers, this work offers advanced solutions to enhancing properties of common building materials to improve and extend their performance.

Emerging Materials and Technologies

Series Editor
Boris I. Kharissov

The *Emerging Materials and Technologies* series is devoted to highlighting publications centered on emerging advanced materials and novel technologies. Attention is paid to those newly discovered or applied materials with potential to solve pressing societal problems and improve quality of life, corresponding to environmental protection, medicine, communications, energy, transportation, advanced manufacturing, and related areas.

The series takes into account that, under present strong demands for energy, material, and cost savings, as well as heavy contamination problems and worldwide pandemic conditions, the area of emerging materials and related scalable technologies is a highly interdisciplinary field, with the need for researchers, professionals, and academics across the spectrum of engineering and technological disciplines. The main objective of this book series is to attract more attention to these materials and technologies and invite conversation among the international R&D community.

Sustainable Nanomaterials for the Construction Industry
Ghasan Fahim Huseien and Kwok Wei Shah

4D Imaging to 4D Printing
Biomedical Applications
Edited by Rupinder Singh

Emerging Nanomaterials for Catalysis and Sensor Applications
Edited by Anitha Varghese and Gurumurthy Hegde

Advanced Materials for a Sustainable Environment
Development Strategies and Applications
Edited by Naveen Kumar and Peter Ramashadi Makgwane

Nanomaterials from Renewable Resources for Emerging Applications
Edited by Sandeep S. Ahankari, Amar K. Mohanty, and Manjusri Misra

Multifunctional Polymeric Foams
Advancements and Innovative Approaches
Edited by Soney C. George and Resmi B.P.

For more information about this series, please visit: www.routledge.com/Emerging-Materials-and-Technologies/book-series/CRCEMT

Sustainable Nanomaterials for the Construction Industry

Ghasan Fahim Huseien and Kwok Wei Shah

CRC Press
Taylor & Francis Group
Boca Raton London New York

CRC Press is an imprint of the
Taylor & Francis Group, an Informa business

First edition published 2023
by CRC Press
6000 Broken Sound Parkway NW, Suite 300, Boca Raton, FL 33487-2742

and by CRC Press
4 Park Square, Milton Park, Abingdon, Oxon, OX14 4RN

CRC Press is an imprint of Taylor & Francis Group, LLC

© 2023 Ghasan Fahim Huseien and Kwok Wei Shah

ISBN: 978-1-032-25090-8 (hbk)
ISBN: 978-1-032-25091-5 (pbk)
ISBN: 978-1-003-28150-4 (ebk)

DOI: 10.1201/9781003281504

Typeset in Times
by codeMantra

Contents

Preface...ix
Author Biographies xi

Chapter 1 Nanomaterials-Based Sustainable Pigments......................................1

    1.1   Introduction 1
    1.2   Core–Shell NP: Synthesis Approach and Importance 2
    1.3   Core–Shell Synthesis Methods..4
    1.4   Materials-Based Shell Part... 10
    1.5   Efficiency and Test Methods 10
    1.6   Applications of Core–Shell Pigments 15
    1.7   Summary 16
    References 17

Chapter 2 Modification of Cement-Based Materials with Nanoparticles...........23

    2.1   Introduction 23
    2.2   Nanotechnology and Nanomaterials 25
    2.3   Nanomaterials-Modified Cement Binder 27
    2.4   Sustainability Performance 33
    2.5   The Interfacial Transition Zone (ITZ).....................................35
    2.6   Carbon Nanomaterials Applied in Cementitious
            Composites 41
    2.7   Nanomaterials-Based Geopolymer Concrete 44
    2.8   Effects of Nanomaterials on Alkali-Activated Binders........... 47
    2.9   Summary 52
    References 52

Chapter 3 Nano-Enhanced Phase-Change Materials 67

    3.1   Introduction 67
    3.2   Nano-Metal Enhancer 68
    3.3   Nano-Metal Oxide Enhancer... 73
    3.4   Nanocarbon Enhancer 77
    3.5   Summary 83
    References 84

Chapter 4 Preparation and Properties of Nanopolymer Advanced
                Composites 87

    4.1   Introduction 87
    4.2   Compatibilisation in Polymer Nanocomposites 88

        4.2.1   In Situ Polymerisation 88

        4.2.2   Solution Blending 89

        4.2.3   Melt Blending 90

  4.3   Nanopolymer Fibre-Reinforced Composites 91

  4.4   Nanopolymer Fibre-Reinforced Sandwich Composites 93

  4.5   Nanopolymers and Their Applications 94

  4.6   Environmental Applications 95

        4.6.1   Heavy Ion Removal 96

        4.6.2   Solar Energy 97

  4.7   Polymeric Nanofibres as Sensors 97

  4.8   Summary 99

  References 99

Chapter 5   Nanotechnology-Based Smart Glass Materials 103

  5.1   Introduction 103

  5.2   Nanotechnology 104

  5.3   Self-Cleaning Glass 106

  5.4   Hydrophilic Coating 107

  5.5   Anti-Reflective Coating 108

  5.6   Photocatalytic Activity of TiO_2 109

  5.7   Fabrication of Self-Cleaning Glass 110

  5.8 SiO_2–TiO_2 Coating 110

  5.9   Nanomaterial-Based Solar Cool Coatings 111

        5.9.1   Metal-Based Nanoadditives 112

        5.9.2   Metal Oxide-Based Nanoadditives 112

        5.9.3   Absorption-Based Nanoadditives 115

        5.9.4   Metalloid-Based Nanoadditives 115

  5.10   Key of Building Applications 116

        5.10.1 Gold (Au) 116

        5.10.2 Zinc Oxide (ZnO) and Aluminium Zinc
                Oxide (AZO) 116

        5.10.3 Indium Tin Oxide (ITO) and Antimony Tin
                Oxide (ATO) 117

        5.10.4 Vanadium Oxide (VO_2) 119

        5.10.5 Tungsten Oxide (WO_3) and Alkali Metal-Doped
                Tungsten Oxide ($AxWO_3$) 120

  5.11   Summary 122

  References 122

Chapter 6   Air Nano Purification 125

  6.1   Introduction 125

  6.2   Metallic Oxides 127

        6.2.1   Titanium Dioxide (TiO_2) 127

        6.2.2   Zinc Oxide 132

	6.2.3	Nickel Oxide	133
	6.2.4	Tungsten Trioxide	133
	6.2.5	Manganese Oxide	134
	6.2.6	Bi-Based Compounds	134
	6.2.7	Ag-Based Compounds	134
	6.2.8	Platinum-Supported Material	135
	6.2.9	Iridium Particles	135
6.3	Carbon-Based Materials		136
	6.3.1	Carbon-Based	136
	6.3.2	Graphene and Graphene Oxide (GO)	136
6.4	Applications on Buildings		137
	6.4.1	Indoor Air Treatment	137
	6.4.2	Coating	138
	6.4.3	Paints	139
	6.4.4	Construction Materials	139
6.5	Summary		140
References			140
Index			147

Preface

Globally, environmentally sustainable construction materials with high performance have been in great demand by the construction industry. Recently, the growth of nanotechnology and the accessibility of nanomaterials suited for construction usage including nano-alumina, nanosilica, nano-kaolin, nano-titanium, etc. have enhanced remarkably the properties of construction materials. The nanoparticles being the linkages among the bulk structures as well as the atomic and molecular structures, their emergent novel characteristics mainly depend on the individual components. Consequently, the overall attributes of the nanoparticles are appreciably different than the one that exist at larger dimensions. Regardless of their nature, the most important properties that lead to the extensive uses of these nanoparticles are related to their enlarged surface area, emergent optical traits due to the quantum size effects, improved absorbance, uniformities and surface functionalisations. Especially, the effect of quantum confinement in the nanoparticles leads to the emergence of the spontaneous semiconducting, conducting or insulating properties for the adjacent particles, improved stabilities, chemical and physical behaviours.

This book deals with a review of past research works on nanomaterials-based construction materials. Several applications of nanoparticle materials in construction industry are widely discussed in this book's chapters. In this book, we focussed on benefits of nanomaterials for pigment, cement binder, phase-change materials, polymer composite, building glass and air purification enhancement. Effect of nanomaterials type, particle size and content on sustainability construction materials has been widely discussed.

Ghasan Fahim Huseien
Kwok Wei Shah

Author Biographies

Dr. Ghasan Fahim Huseien Huseien is a Research Associate at the Department of the Build Environment, School of Design and Environment, National University of Singapore, Singapore. He has over 5 years of Applied Research and Development as well as he has up to 12 years' experience in manufacturing smart materials for sustainable building and smart cities. He has expertise in Advanced Sustainable Construction Materials covering Civil Engineering, and Environmental Sciences and Engineering. He authored and co-authored +135 publications and technical reports, 6 books and 22 book chapters and participated in +35 national and international conferences/workshops. His past experience in projects including application of nanotechnology in construction and building materials, self-healing technology, geopolymer as sustainable and eco-friendly repair materials in construction industry.

Prof. Dr. Kwok Wei Shah Professor Shah is presently Assistant Professor, and Deputy Program Director under the Deptartment of Building, School of Design and Environment, National University Singapore. He is Advisory Board member of Vietnam Green Building Council and sits on VGBC Education Committee. He lectures for REHDA GreenRE in Malaysia and Visiting Fellow of University Technology of Malaysia, UTM. He is Visiting Professor at Tianjin University of Technology, China. He is appointed BCA Ambassador for 3 years period and a member of SPRING and SGBC technical committees. He served as Technical Consultant for Ascendas Services Pte Ltd, Chief Technical Advisor for Bronx Culture Pte Ltd.

Dr. Shah's research interest is on nanotechnology and nanomaterials for green building applications. Dr. Shah has done outstanding research work on a novel low-cost, high-volume aqueous silica-coating technique has been granted a US Patent (US 20130196057 A1). His research paper published by *Nanoscale* (*Nanoscale*, Impact Factor = 6.739, doi:10.1039/C4NR03306J) on "Noble Metal Nanoparticles Coated with Silica by a Simple Process That Does Not Employ Alcohol" was highlighted by popular online science magazines such as ScienceDaily, Physorg and A*STAR website. Separately, Dr. Shah's research on microencapsulated phase-change materials enhanced by highly thermal conductive nanowires (*Journal of Materials Chemistry A*, doi:10.1039/C3TA14550F, Impact Factor = 6.626) led to the development of "M-KOOL" phase-change cooling technology, which was featured on Physorg, Channel News Asia, *Straits Times*, *Business Times*, TODAY, The Star Online and Lianhe Wanbao. So far, Dr. Shah's achievements include 3 first-authored papers, 9 co-authored papers, 1 book chapter, 12 patents disclosures and 1 commercial licensing.

1 Nanomaterials-Based Sustainable Pigments

1.1 INTRODUCTION

Synthetic-coloured pigments that have been launched in the market in the past few years resulted in more extensive scientific research focused on this area. Typical applications of pigments are varnishes, paints, plastics and textiles, printing inks, building materials and rubber, ceramic glazes and leather decoration [1–3]. The definition of pigment durability is its ability of resisting weathering processes and negating deteriorating when placed in an external environment [4].

Recent research has shown that efficient energy consumption and environmental protection measures are deemed significant [5]. To address this issue, the production of both sustainable and durable pigments has become the fundamental requirement within the construction industry. A myriad of methods have been applied in order to increase the pigments' durability, and the most significant is known as the core–shell method [6–9].

In recent years, there has been a surge in the development of various chemical synthesis techniques. Such research has found that multi-component materials possess diverse compositions and structure. These attributes signify remarkable property type, and they are applicable in various fields [10–13]. There is even more research being conducted on their distinctive core–shell structure.

There are many advantages of the core–shell structure in comparison to other types of composite materials. One such advantage is their ability to generate or increase the strength of new chemical and physical capabilities, enable maintenance on structural integrity, deter the core from breaking up to large particles and ascertain dispersion effectively. In addition, they also provide conventional multifunctional compositions and structure with other advantages. Moreover, a synergetic effect between the shells and cores would even extend the performance further [14].

Science and technology field has been attentive on the phenomena of materials that are derived from the core–shell properties because the properties can be finely customised [1,15,16]. A shell domain cloaks a core structural domain within each of the core or shell particles. Materials that possess core or shell particles include inorganic solids, metals and polymers. There is no difficulty in modifying characteristics such as size and structures as well as the particles' composition in order to further customise their optical, magnetic, mechanical, thermal, electrical, catalytic and electro-optical properties.

Core or shell morphology could be applied to produce hollow spheres and minimise costs on precious materials. Thus, the materials with the reduced core costs can be coated to precious materials [17,18]. Particles with size less than 0.1 µm diameter are classified as nanoparticles (NPs) and have been gathering much attention in

DOI: 10.1201/9781003281504-1

1

research in the past few years. Essentially, NPs are smart materials with exclusive properties.

Applications using NPs have more advantages compared to materials that have larger surface-to-volume ratio such as microscale, macroscale and bulk materials [19,20]. Due to increased research on NPs development, it is now possible to synthesise NPs in symmetrical shapes including spherical, prism, hexagon, cube, wire, tube and rod [21–23]. Despite this achievement, the bulk of the research is still at early stage in terms of exploring the possible shapes that can be synthesised.

Recent research identified the ease of production method for nanoparticles that are non-spherical [24–26]. However, it must be stressed that NPs' properties are dependent on their actual shape and size. Such properties that are dependent on particle size include temperature barrier, magnetic saturation and permanent magnetisation. Furthermore, coactivity of nanocrystals is dependent on the particle's shape, as it has a direct influence on the surface anisotropy [27].

Rapid advancements are made in nanotechnology resulting in the founding of core–shell NPs, which is a leading functional material. This has attracted even more research conducted on various functional compositions of core–shell NPs where it could be applied in a variety of areas such as optics, catalysis, biomedicine, electronics and medicines [28]. Core–shell NPs possess beneficial physiochemical properties that are exclusive, and this attribute has garnered a lot of researchers' attention. The primary advantages of core–shell NPs are that it could increase protection level, encapsulation and controlled release [23,29].

The discovery of a variety of core–shell NPs leads to its applications to a variety of situations. The difficulty, however, is to identify the individual type core–shell NPs that are applicable to the respective industries due to its multitude of types. Research on core–shell NP pigments is focussing on core–shell materials and their production methods, distinctive properties and applications. This chapter discusses the primary feature and properties of core–shell NPs, which include their fabrication methods, inorganic materials and typical applications.

The chapter begins with a brief outline of the various methods of production along with a discussion of the various classifications of the core–shell materials that are already in use. Next, an outline will be presented pertaining to the primary new fabrication methods of the core–shell NP pigments within all research fields. The final chapter discusses the application of core–shell NPs within paints designed for roads.

1.2 CORE–SHELL NP: SYNTHESIS APPROACH AND IMPORTANCE

Nanotechnology utilises biological, engineering and chemical methods to customise materials at their atomic level. NPs can be synthesised using various biological, chemical, physical and hybrid methods (Figure 1.1). There is a further classification for such synthetisation: "top-down" and "bottom-up". The top-down approach incorporates conventional workshops with microfabrication methods in addition to the equipment that are externally controlled, which is used to mill, cut, shape and mould the materials according to the required shape.

Lithographic and mechanical techniques are conventional top-down approaches. Lithographic techniques involve electron or ion beam, UV, scan probing, optical

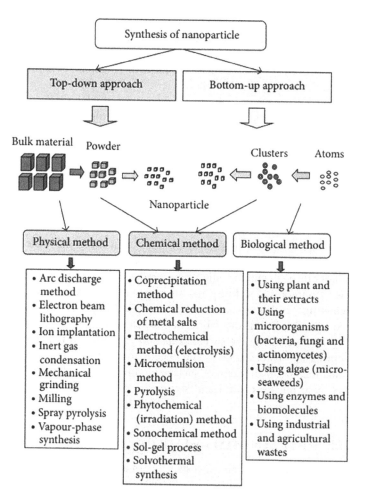

FIGURE 1.1 Synthesis of nanoparticles using different approaches [38].

near field scanning and laser-beam processing. Meanwhile, mechanical techniques involve machines that grind, cut and polish the materials according to the required specifications [30–33].

Bottom-up techniques, on the other hand, assemble the desired form on the chemical composition down to the molecular level. Examples of typical bottom-up techniques include chemical vapour deposition, laser-induced assembly, chemical synthesis, self-assembly, colloidal aggregation, and film deposition and growth [34,35].

There are both advantages and disadvantages of using both approaches, and no approach has more advantages than the other. The main advantage of bottom-up approach, however, is its cost-effectiveness, and it could fabricate significantly smaller particles in comparison to the top down approach. This is due to its precision as the product is produced by assembling it down to molecular level; thus it is possible to have total control and almost have no energy loss on the entire production process. The synthesis of core–shell NPs necessitates total control in order to coat the shell materials uniformly as the particles are formed. Therefore, bottom-up approach is more suitable for such synthetisation.

Hybrid approach involves both the aforementioned techniques. An instance is where the core particles can be produced via the top-down approach, while bottom-up approach could address the uniformity of the thickness of the shell. It is recommended to apply microemulsion for size and thickness regulation of the shell because water droplets could play the role of nano-reactors. More researchers have been focussing on core–shell due to its suitability to be used extensively in a variety of fields such as electronics, optics, chemistry, biomedicine, medicines and catalysis.

Furthermore, such NPs have high functioning and distinctive properties as different materials can be used as core or shell. The core or shell can be highly customisable by modifying the properties through controlling the materials or the core-to-shell ratio [36]. It is also possible to modify the core particles' reactivity and thermal stability through making adjustments to the shell coating material. This will ultimately lead to improved stability and dispersion of the core particles.

This means that each particle can have exclusive properties depending on the materials being used during the fabrication. Such technique is renowned due to its capability of customising the surface function according to the environment by applying appropriate materials [37]. The benefits of coating core particles include improved functionality, facilitation of surface modifications, stability, core particles dispersibility, core release control and significant decrease in the use of precious material.

1.3   CORE–SHELL SYNTHESIS METHODS

Core–shell particles, as the name suggests, contain a shell and a core. A shell could be produced using the same or different materials that are used for the core [39–41]. Figure 1.2 shows the variety of core–shell particles where the colours are used to differentiate between single sphere as shown in Figure 1.2a or multiple spheres that are smaller in size as seen in Figure 1.2b. Furthermore, a shell may be hollow with one small sphere inside that looks similar to a yolk–shell structure (Figure 1.2c) [42].

There are three forms of shell structure, a continuous layer as shown in Figure 1.2a–c: a larger core sphere that contains many smaller spheres as shown in Figure 1.2d and e or simply a collection of core spheres as shown in Figure 1.2f [43]. In order to extend the intricacy of core–shell structure, it is possible to insert smaller spheres into the shell (Figure 1.2g) [44]. This process of inserting smaller shells could also be done through multiple shells as shown in Figure 1.2h [45,46]. Core–shell NPs can be synthesised by physical or chemical approaches, which include deposition of chemical/physical vapour and wet chemistry.

The process of core–shell particles synthetisation normally involves different stages. The most fundamental is to firstly synthesise the core particles forming the shell onto the particle. This method, however, is dependent on the core type and shell materials being utilised [44]. The main goal of producing core–shell particles is to

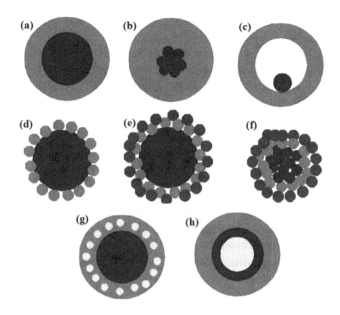

FIGURE 1.2   Schematic representation of different types of core–shell particles [39].

utilise the suitable materials and structures. Subsequently, the desirable attributes can be achieved such as active particles stabilisation, creating biocompatible properties and synergy effect [47].

Many industries from various fields utilise a basic nanomaterial for synthesising core–shell NPs. The main features to determine prior to synthesising core–shell NPs are the speed, simplicity and cost-effectiveness in addition to causing minimal damage to the environment. Many methods have been established to meet the aforementioned requirements such as electrochemical dealloying, sol-gel, sonochemical processing, microwave synthesis, multi-step reduction, micro emulsion, epitaxial growth and the Stöber method. The hybrid method involves more than one of the aforementioned methods.

A typical method of producing core–shell NPs is the sol-gel synthesis method. Sol-gel method has been recently developed, which prepares NPs in order to establish additional control during the reaction processes during the synthesising of materials that are solid. It is easy to obtain homogenous multi-component systems especially homogenously mixed oxides where it could be produced through mixing solutions containing molecular precursors. Essentially, sol-gel method forms solid materials through the use of small molecules and are typically used to synthesise metal oxides (most commonly silicon oxides (SiO_2) and titanium oxides (TiO_2)).

Synthesising metal oxides involves converting monomer into a colloidal solution (sol). The solution is the precursor to be used for combination of networks including discrete particles or network polymers. An example of frequently used precursor is metal alkoxides. Sol is produced when a chemical reaction occurs and eventually produces a diphasic substance that has a property similar to gel, which is both liquid and solid. These phases may be either discrete particles or continuous polymer networks formed during morphologies.

Turning colloid into properties such as gel necessitates removing a large volume of liquids in the events that the volume of particle density is significantly low. One of the simplest methods to achieve this is to have adequate time for sedimentation to occur before disposing the remaining liquid. In contrast, centrifugation could be applied to accelerate phase separation. Sol-gel is a more common wet chemical method used to synthesise core–shell NPs [48–50].

Microemulsions are a mixture of isotropic liquid that is made up of surfactant, oil, water and more commonly, cosurfactant. It has a clear appearance and a stable thermodynamic nature with the presence of salt and other ingredients in liquid form. The appearance of oily substance may be due to the presence of various types of hydrocarbons that are mixed in a complex manner. In comparison to conventional emulsions, microemulsions are synthesised through mixing components and do not necessitate high shear conditions during the production process. The three types of microemulsions are direct (dispersion of oil in water, o/w), reversed (dispersion of water in oil, w/o) and bicontinuous.

Microemulsions belong to the ternary systems. This is when two immiscible substances (water and oil), which form separate layers, react with a surfactant resulting in monolayer formation between the immiscible substances from the surfactant's molecules. During the oil phase, the hydrophobic tails of the surfactant molecules would dissolve. During the liquid phase, however, the hydrophilic head groups would dissolve. Two-step microwave irradiation is the conventional method in rapid synthesisation of gold and silver core–shell bimetallic NPs.

Such a technique necessitates the development of bilayer organic barrier surrounding the core. In turn, this enables a well-defined boundary layer to be established between the core and the shell material. The boundary layer is significant for synthesising a variety of core–shell particles, which are ultimately used in producing customised bimetallic NPs with core–shell structures that are well-defined. The cores of these NPs are triangular or spherical in shape.

A alternative method for producing core–shell materials including nanotubes is high-pressure chemical vapour deposition. Nikolaev et al. [51] found this method for producing a single-wall carbon nanotube. This is achieved using a small amount of $Fe(CO)_5$ to comb CO. The mixture will then be passed to a heat reactor. El-Gendy et al. [52] used this technique to synthesise NPs that are coated with materials including Fe, Co, Ni, FeRu, CoRu, NiRu, NiPt and CoPt. This method allows for the total control of reactors' temperature and pressure.

El-Gendy et al. experiments utilised metal-organic precursors, which are called metallocenes or metals that are rich in carbon. These precursors were inputted into a thermostatic sublimation chamber before releasing argon gas for the purpose of pushing the vapour into the hot zone of the chamber. Firstly, the precursor will break down the NPs within the cooling finger before turning into gas form within the hot zone for supersaturation. Once supersaturation is initiated, the NPs will be nucleated.

Adjustment can be made to the temperature and pressure/temperature within the sublimation chambers and chemical vapour deposition reactor, respectively, in order to achieve the desired degree of supersaturation. When pressure is at high level, collision of the gas atoms increases and the rate of atoms diffusion from the original location decreases. It is worth noting that supersaturation could not occur when diffusion rate is poor. When this occurs, the cooling finger will have deposits of tiny clusters of atoms or individual atoms.

Another investigation [53] has been conducted on the wet chemical technique for synthesising Fe_2O_3 with graphene shells as coating. Oleic acid and 1-octadecene were mixed in a solution before being placed in the reflux reactor to be heated to 320°C for dissolving the iron oleate. The solution is then washed with ethanol and acetone to obtain the iron oxide particles. The Stöber process normally involves preparation of silica (SiO_2) particles [54] that are under total control and uniform in size [55]. These particles have numerous applications within the material science field.

Stöber et al. in 1968 [54] founded the Stöber method, which remains the most renowned wet chemical approach in the synthesisation of NPs [56]. It is a sol-gel process where a molecule, which is normally tetraethylorthosilicate, acts as the precursor when immersed in water. Alcoholic solution is added to form a reaction resulting in the formation of new molecules, which join to form larger objects.

Du et al. [57] use the sol-gel approach to create the necessary reaction to produce SiO_2 shell as a coating for Fe_3O_4 NPs, and evidently, the end product is a core–shell structure. The procedure of synthesising core–shell NPs involves two stages. The first is initiating co-precipitation to obtain Fe_3O_4 NPs and subsequently causing a reaction with tetramethylammonium hydroxide (TMAOH), which results in the formation of a liquid solution that contains the particles. The second is producing SiO_2 through hydrolyzation of tetraethyl orthosilicate (TEOS) in order to limit Fe_3O_4. Figure 1.3 shows Li et al.'s [58] sol-gel approach to produce $ZnSiO_3/ZnO$ core–shell NPs.

The experiment involved two methods: sol-gel and annealing in order to produce core–shell NPs (zinc silicate-zinc oxide ($Zn_2SiO_4@ZnO$)) that has a high bandgap. The start of the procedure is to modify the $Na_2SiO_3/ZnCl_2$ concentrations, thus resulting in $ZnSiO_3$, which forms shells with varied thicknesses before using it to coat ZnO NPs. Next, an annealing temperature is set to a low level at 780°C. Finally, a reaction between the amorphous $ZnSiO_3$ and ZnO occurs; thus, a crystalline Zn_2SiO_4 shell is formed. Chai et al. [48] adopted this technique to synthesise $Fe_3O_4@SiO_2$ NPs consisting of core–shell structure. Their first step was to fabricate Fe_3O_4 NPs through

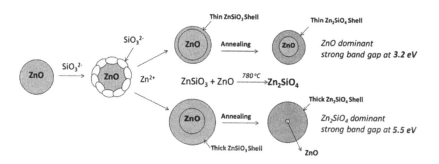

FIGURE 1.3 Sol-gel method for fabrication of core–shell particles.

solvothermal technique. Next, the hydrolyzation of TEOS resulted in SiO_2, which acts as the coating for Fe_3O_4 NPs.

Another example of the two-step reduction technique is Ref. [59], which attempted to synthesise the epitaxial growth of Au@Ni core–shell nanocrystals. The core–shell nanocrystals were synthesised by mixing the hexagonal platelike, decahedral, octahedral, triangular, hexagonal platelike and icosahedral. Subsequently, ethylene glycol (EG) is used for the reduction of $HAuCl_4$, before being placed in a microwave with polyvinylpyrrolidone (PVP), which acts as a polymer surfactant to be heated. The core seeds were produced at this stage, and subsequently, oil bath heating was applied for reducing $Ni(NO_3)_2 \cdot 6H_2O$ in EG with NaOH and PVP. The Ni shells thus will overgrow within the Au core seeds.

Fan et al. [60] conducted a similar technique involving two steps but focussing on seed-mediated growth through the use of Au cores in order to conduct tests on the synthesisation of liquid form of bimetallic core–shell nanocubes. A comprehensive assessment is made upon the heterogeneous core–shell formation on the four common metals which are gold, silver, palladium and platinum. The experiments have identified the following: (i) the general conditions and growth modes to attain conformal epitaxial growth and (ii) heterogenous nucleation and formation of various noble metals.

They were thus able to identify two types of growth for heterogeneous metal shells on gold cores, namely conformal epitaxial growth (Au@Pd and Au@Ag nanocubes) and heterogeneous nucleation and is-land growth (Au@Pt nanospheres). Further findings include two metals with comparable lattice constants where the mismatch should be less than 5%. Similar findings are also concluded by other research where Au@Ag (lattice mismatch, 0.2%), Au@Pd (4.7%) and Pt@Pd (0.85%) [54–56].

Tsuji et al. [61] use another form of synthesisation of Ag@Cu core–shell NPs, which is a one-polyol technique. They have achieved a high yield. The method involves bubbling Ar gas with added reagents ($AgNO_3$ and $Cu(OAc)_2 \cdot H_2O$). The procedure started with a two-step process involving the synthesisation of Ag@Cu particles through $AgNO_3$ reduction in EG. Cu shells were developed by separating the silver cores from $AgNO_3$, before adding $Cu(OAc)_2 \cdot H_2O$. This procedure did not turn out as expected as no Cu@Ag core–shell particles failed to develop but the Cu/Ag bicompartmental particles appeared instead.

The present study implemented various experimental processes combined with different reaction temperatures and heating times in an attempt to produce Ag@Cu particles. The result is that the optimal method for producing Ag@Cu particles is to add two reagents in reverse. To start the process, 8 mL of 15.9 mM Cu (OAc)2. H2O in EG and 8 mL of 477 mM poly(vinylpyrrolidone) (PVP, MW: 55000 monomer units) were placed in EG and the solution was mixed in 100-mL three-necked flask.. A 100 mL three-neck flask was used for mixing the solution. Ar was bubbled for 10 minutes at room temperature for the purpose of total removal of oxygen from the solution and subsequently being soaked in an oil bath at 180°C. The solution is continued to be bubbled while the temperature is being raised to 175°C. Afterwards, the reagent solution will be added with 2 mL of 15.7 mM $AgNO_3$, which will be left for 20 minutes while maintaining the temperature at 175°C. The final concentration's result was 7.0 and 212 mM for Cu (OAc)$_2$•H$_2$O, 1.7 mM for $AgNO_3$ and PVP, respectively. Further investigation was conducted by varying the reaction time on the reagent solution to observe the growth process of Ag@Cu.

Chae et al. [48] attempted to produce $Fe_3O_4@SiO_2$ through the use of customised Stöber method where a sol-gel method was used. This necessitates the ultrasonication of the solution of $4\,g$ Fe_3O_4 particles and additional TEOS to be raised from 4 to $40\,mL$. After this, the emulsion was inserted into a mixture containing $50\,mL$ ethanol and $12\,mL$ $NH_3\,H_2O$. The reaction solution will be stirred at $400\,rpm$ at room temperature for 4 hours before separating the core–shell structured $Fe_3O_4@SiO_2$ NPs through the use of centrifugation. This entire process of $Fe_3O_4@SiO_2$ synthesisation is shown in Figure 1.4.

Another similar research conducted by Sharma et al. [62] has shown that it is possible to fabricate core–shell particles through precipitation without the need for a surfactant. Despite this, the research included the outcome if different surfactants were used as well as the concentrations of core–shell particles as they were formed. The research in Ref. [62] used both anionic and non-ionic surfactants. The nanoTiO$_2$ is developed as a form of shell through the use of fly ash. The primary reason for the use of surfactants is to strengthen the adhesion of the nano titania shells to fly ash core.

Therefore, different types of surfactants were used to test the strength of TiO_2 adhesion onto fly ash. As previously mentioned, both anionic and non-ionic surfactants were used as well as one test being conducted with no surfactant. When anionic surfactant was used, the resulting particles have remarkable pigment properties and reflectance within the near-infrared area. This means that these pigments are suitable for cool coating applications. A solution of 70% ethanol is added with the following in sequence, fly ash, anionic (SDS) or non-ionic surfactant (TX-100) and finally titanium isopropoxide. Finally, the solution is stirred for 2 hours before being dried at $50°C–60°C$. The final obtained substance is in a powder form.

Another research conducted in Ref. [63] attempted to produce PUA hybrid emulsion PA/PU with a ratio of 20/80 through the use of semi-batch emulsion. The equipment setup for this research was digital thermometer, $250\,mL$ four-neck glass flask containing a reflux condenser, mechanical stirrer and nitrogen gas inlet. The pre-emulsion was prepared by dissolving $2.0\,g$ per $100\,g$ of acrylic and PU content into the water before gradually adding $5.0\,g$ of MMA, $5.0\,g$ of BA and $0.015\,g$ of AA (0.15 wt% of the overall MMA and BA weight). The solution was then stirred before mixing a further 0.5 hour.

The objective is to obtain $111.3\,g$ PU emulsion dispersion and 10% monomers from the reactor vessel; thus the temperature is set at $80°C$ while the contents were stirred. Next, $0.4\,g$ of KPS per $100\,g$ acrylic monomers, which is made up of 10%,

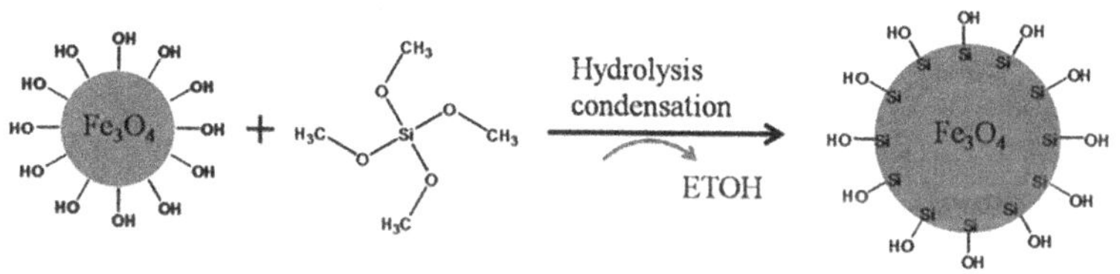

FIGURE 1.4   The Stöber method for $Fe_3O_4@SiO_2$ nanoparticles synthesis [48].

was added and continuously stirred for 30 minutes. Subsequently, the temperature was increased to 5°C, and at the same time, the leftover monomer pre-emulsion and initiator solution was flowed into the task for 240 minutes at a constant flow rate. After 240 minutes, the solution was left at 85°C for 0.5 hour before being stirred once again and wait for the temperature to drop. Lastly, the pH was maintained at the desirable range after adding $NaHCO_3$.

1.4 MATERIALS-BASED SHELL PART

Several materials such as metals and biomolecules are used to create core–shell NPs. There are two components, the central core and an alternative core, which is the shell. Core-shell nanostructures have excellent thermal and chemical stabilities, high levels of solubility, low toxicity and greater permeability to certain target cells, which makes them very beneficial for use in a variety of sectors. Such properties enable them to have vast potential in many sectors. Furthermore, micro–nano-scale core–shell particles have attributes that are exclusive and unique to themselves when compared with other particles.

They effectively bring together the properties of the materials used in the core and shell, as well as smart properties generated through their materials. In the past few years, there have been an increased research interest in core–shell structures production [64]. This is particularly true within the pigment industry due to the high range of applications of core–shell materials in order to increase pigments' durability. The core–shell materials could be made of both organic and inorganic materials.

An example of this is an experiment conducted by Cao et al. [65] to develop hybrid pigments with inorganic–organic structure by a mixture of precipitated SiO_2 and TiO_2. Furthermore, dye core@silica shell structure is developed through the use of mesoporous soft template synthesis [66]. This section explores the possibility of using inorganic materials to produce core–shell materials with a focus on SiO_2 and TiO_2.

1.5 EFFICIENCY AND TEST METHODS

The following tests are examples that are capable of testing the synthesisation method of the core–shell NPs: SEM, LC-MS, XPS, FTIR, XRD, TEM, BET, ultraviolet–visible spectroscopy, Raman spectrum and near-infrared reflectance, and photoluminescence spectroscopy techniques [40,41,50,67–71]. For instance, assessment of morphology, chromaticity and structure of α-Fe_2O_3@SiO_2-fabricated pigments can be tested with SEM, TEM, FTIR, XPS and XRD [29]. Figure 1.5a shows the patterns of XRD, which consist of the pigments belonging to α-Fe_2O_3@SiO_2 NPs, α-Fe_2O_3@SiO_2 and α-Fe_2O_3.

The formation of the core–shell structures results in the diffraction peak of α-Fe_2O_3@SiO_2 particles in wide ($2\theta = 15°$–$25°$). This indicates the presence of amorphous SiO_2. Furthermore, 1000°C calcination changes the diffraction peak to around 22°. This results in pigment that is reddish in colour indicating that the amorphous shell has entered into a cristobalite phase. In addition, the formation of the core–shell structure weakens the α-Fe_2O_3 diffraction peak.

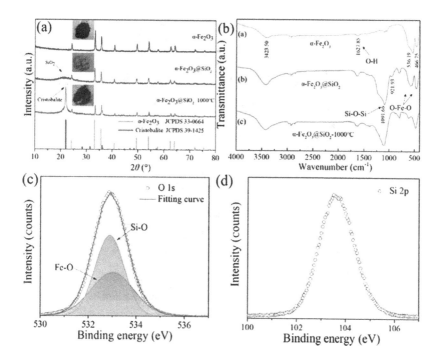

FIGURE 1.5 (a) XRD patterns. (b) FTIR spectra of different samples of prepared pigments; high-resolution XPS spectra of (c) O 1s. (d) Si 2p for α-Fe$_2$O$_3$@SiO$_2$ pigments calcined at 1000°C [29].

Figure 1.5b shows the FTIR result of the reddish pigments of α-Fe$_2$O$_3$, α-Fe$_2$O$_3$@ SiO$_2$ NPs and Fe$_2$O$_3$@SiO$_2$. The hydroxyl (−OH) stretching results in bands at 3423.50 and 1627.85 cm^{-1}, which peaks at 536.19 and 466.75 cm^{-1}, which has a correlation to the O−Fe−O bands of α-Fe$_2$O$_3$. The band is at 1091.66 and 470 cm^{-1} during the covering of α-Fe$_2$O$_3$ in SiO$_2$ as a result of the bending and stretching of O−Si−O. This is an indication that a coat is forming on the α-Fe$_2$O$_3$ surface.

In addition, the calcination of O−Si−O bond will increase its strength as well as improve the core and shell interactions. Figure 1.5c and d shows the assessment results of the reddish pigments through the use of XPS. Figure 1.5c shows that Fe-O bonds and Si-O bonds are in the O 1s pigment as evidenced by the high-resolution XPS spectrum. At the same time, a band exists at 103.5 eV in the Si 2p XPS spectrum, which is expected in pure silica.

Conversely, Li et al. [50] conducted an analysis on the synthesised γ-Ce$_2$S$_3$@SiO$_2$ core–shell materials by means of a TEM test. Figure 1.6 shows the silica shell being formed at various coating times. A clear layer covers the γ-Ce$_2$S$_3$, but it is not found on the samples that are not coated, which is in accordance with the SEM analysis. Figure 1.6b–d shows a correlation between the increasing thickness of the coating layer and increasing coating times. This is demonstratable with an example where when the particles were coated once, twice and three times, its thickness was 70, 100

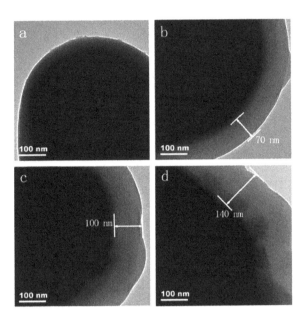

FIGURE 1.6 TEM images of (a) uncoated γ-Ce_2S_3 and (b) once-, (c) twice-, and (d) thrice-coated γ-Ce_2S_3@SiO_2 core–shell particles [50].

and 140 nm, respectively. This shows that it is possible to control the coating thickness through number of coatings being applied.

Liu et al. [72] used four types of tests (FTIR, TEM, XRD and EDS) for the assessment of the morphology of fabricated Ce_2S_3@SiO_2 samples. The first step is to assess the SiO_2 thickness used to coat γ-Ce_2S_3. This is done through the TEM test. Figure 1.7 shows the different amounts of volume ratio of water/ethanol being used for the preparation of uncoated γ-Ce_2S_3 pigments and SiO_2 xerogel-coated γ-Ce_2S_3. Figure 1.7a presents the surface, which is deposited with irregularly large chunks with small particle upon an uncoated γ-Ce_2S_3 pigments.

Furthermore, detection of Zn signals occurred within the energy-dispersive spectroscopy (EDS). This means that the colour stability of the uncoated γ-Ce_2S_3 pigments can be controlled by using ZnO. Another advantage of the application of these pigments is its low H_2S emissions. Figure 1.7b–d shows the presence of core–shell structures within all the pigment particles during coating. Simultaneously, Si signal is detected as shown in Figure 1.7e,f. This proves that SiO_2 xerogel constitutes the coating layer that is formed on the γ-Ce_2S_3 surface.

Figure 1.7b and c, on the other hand, shows that the application of water/ethanol volume ratio of 15/105 (48 nm) and 20/100 (60 nm) results in a moderately uniform shell size. However, Figure 1.7d shows that when the ratio is adjusted to 25/95, the thickness of the shell is no longer uniform. The main reason for this is as water volume has risen, it will accelerate TWOS hydrolysis. This means that during the coating process, shell thickness is no longer in uniform because of the competition between surface and silica nuclei.

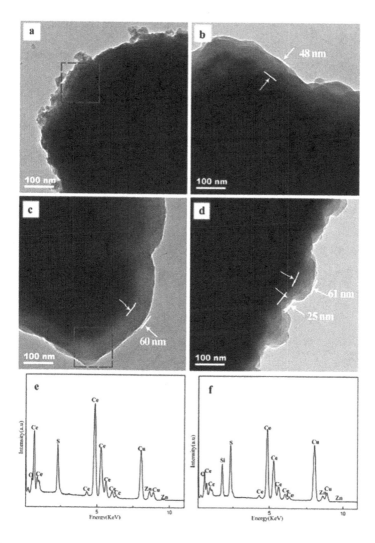

FIGURE 1.7 TEM images and EDS patterns of SiO$_2$ xerogel-coated Ce$_2$S$_3$ prepared with different water-to-ethanol ratios: (a) S0, (b) S1, (c) S2, (d) S3, (e) EDS spectra of S0 and (f) EDS spectra of S2 [72].

Figure 1.8a,b shows the reflective curves that were founded by the research conducted by Sadeghi-Niaraki et al. [73] for the following: CT, CFT2, CFT4, CFT5 and CF. CT sample shows that reflectivity of all wavelengths was improved. This is due to the increase in the crystallinity as it enters the rutile phase. After the calcination, the reflectance value has increased as the sample experiences crystallisation. Figure 1.8c shows that the presence of Fe$_2$O$_3$ results in darker hues within the samples in addition to NIR reflectance being reduced. The NIR solar reflectance for the samples is CT (76%), CFT2 (73%), CFT4 (68.8%), CFT5 (68.4%) and CF (39.3%). Figure 1.8d shows the IR reflectance process within Fe$_2$O$_3$-TiO$_2$ and Fe$_2$O$_3$ particles.

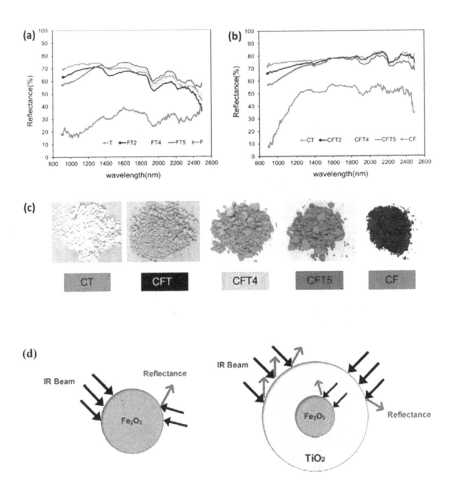

FIGURE 1.8 Reflectance spectra of (a) T, FT2, FT4, FT5 and F samples; (b) CT, CFT2, CFT4, CFT5 and CF samples. (c) Photographs of CT, CFT2, CFT4, CFT5 and CF samples. (d) Proposed mechanism of IR reflectance in Fe_2O_3 and $Fe_2O_3@TiO_2$ composites [73].

In Ref. [74], the authors tested the high temperature tolerance of the red pigments with $Ce_2S_3@SiO_2$-based core–shell. Figure 1.9 shows the XRD patterns related to the γ-$Ce_2S_3@c$-SiO_2 samples, where their production was subjected to various calcination temperatures. There is an absence of the commonly found SiO_2 diffraction peaks when the calcination temperatures occur at the range from 1100°C to 1150°C.

However, the diffraction peaks occurred during the γ-Ce_2S_3 crystalline phase. This means that SiO_2 failed to crystallise. However, c-SiO_2 diffraction peak initiated as the temperature reached 1200°C. This suggests that SiO_2 will only crystallise within Ar gas atmosphere when temperature reaches 1200°C. c-SiO_2 diffraction peak's intensity remains approximately at a constant level as temperature is further raised to 1250°C. Therefore, c-SiO_2 is prone to crystallisation when two conditions are met: (i) it is within Ar gas atmosphere and (ii) temperature to be at least 1200°C.

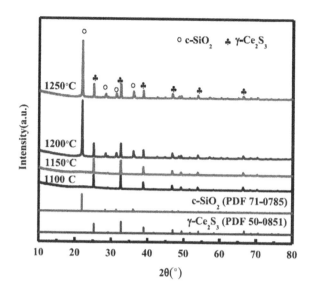

FIGURE 1.9 XRD patterns of the γ-Ce₂S₃@c-SiO₂ samples at different sintering temperatures in Ar gas atmosphere [74].

Li et al. [75] concluded the research by conducting an analysis on the γ-Ce₂S₃ red pigments' resistance through XRD test.

1.6 APPLICATIONS OF CORE–SHELL PIGMENTS

Pigments can be used for decoration or delineation purposes in public streets, thus improving both the aesthetics and public safety. Infrastructures that are well-built and well-planned motivates individuals to use them for recreational activities such as walking, cycling or ease of access for personal mobility devices (PMDs). If individuals are more willing to do the aforementioned activities, they are less likely to use their cars for short destinations, which in return contributes to their overall healthy lifestyle. Figure 1.10 shows a variety of red pigments being applied on Singapore roads for pedestrian use.

Despite advancements in the comprehension of the causes and effects of material failure, it remains a major concern in the entire construction industry. Exterior durability is typically enhanced through the use of high-performance coatings. Pigments are chosen for both the desired colour and performance [76]. The paint industry would exclusively use high-quality pigments. It is important for these pigments' particles to be uniform in size as it could have an effect on the paint's attributes such as lightening capacity, hiding power, tinting strength and gloss. Furthermore, it is mandatory to apply nanoscale pigment particles in luminescent materials for the purpose of UV coatings and colouring.

There is a higher desirability for coloured asphalt and red concrete in comparison to traditional materials as the former has better aesthetics from the viewpoint in

FIGURE 1.10   Red pigments applied on Singapore's cyclist and pedestrian roads.

architecture design [77]. The past few years have seen major development in nano-materials and nanotechnology. This has made synthesising core–shell NPs possible, which also contributed to developing pigments that are sustainable yet with higher colour stability as well as able to tolerate harshness. The development of the pigments with increased durability has led to an increase of potential applications in applying colours onto concrete and asphalt. This leads to further development within the architect industry where they have the options to apply colours that have higher stability and higher tolerance to abrasion. Such development indeed could be combined with the aesthetic and decorative aspects of conventional concrete thus forming an additional material that is attractive.

1.7 SUMMARY

The construction industry has seen increased applicability of the development of sustainable pigments through core–shell and nanotechnologies. The following conclusions are made based on the in-depth and relevant literature overview of nanotechnology-based core–shell pigments:

i. A new class of hybrid and core–shell NPs can be developed due to the advent of the manipulation technique of particle structures at nanoscale level.
ii. There has been successful fabrication of a vast number of core–shell nano-structures through the use of various strategies; as a result, techniques that

have seen a fast-paced advancement include synthetic chemistry, device setup, and colloid and interfacial science.

iii. The objectives of all the methods during the fabrication process for the NPs are for the method itself to be effective and environmentally friendly while at the same time, be able to obtain the desirable features including composition, size control, architecture and properties through the rational molecular design and materials preparation.

iv. The pigments' durability is improved due to the application of core–shell NPs. Furthermore, it is also a part of sustainable materials that have many uses.

v. The highest recommended materials for shells during synthesisation are SiO_2 and TiO_2.

vi. Measuring and evaluation of synthesised core–shell nanostructures could be done using SEM, TEM, FTIR, XRD and EDX. Environmentally friendly materials produced through synthesising methods that are sustainable can address the ongoing pollution challenges. At the same time, such methods involving green chemistry are also economically viable.

REFERENCES

1. Lin, C., et al. A facile synthesis and characterization of monodisperse spherical pigment particles with a core/shell structure. *Advanced Functional Materials*, 2007, **17**(9): pp. 1459–1465.
2. Torres-Cavanillas, R., et al. Downsizing of robust Fe-triazole@ SiO_2 spin-crossover nanoparticles with ultrathin shells. *Dalton Transactions*, 2019, **48**(41): pp. 15465–15469.
3. Shah, K.W., G.F. Huseien, and H.W. Kua. A state-of-the-art review on core–shell pigments nanostructure preparation and test methods. In: *Micro*, 2021. Multidisciplinary Digital Publishing Institute, Switzerland, pp. 55–85.
4. Alexander, M., J. Mackechnie, and W. Yam. Carbonation of concrete bridge structures in three South African localities. *Cement and Concrete Composites*, 2007, **29**(10): pp. 750–759.
5. Sadeghi-Niaraki, S., et al. Preparation of (Fe, Cr)$_2$O$_3$@ TiO_2 cool pigments for energy saving applications. *Journal of Alloys and Compounds*, 2019, **779**: pp. 367–379.
6. Soranakom, P., et al. Effect of surfactant concentration on the formation of Fe_2O_3@ SiO_2 NIR-reflective red pigments. *Ceramics International*, 2021, **47**(9): pp. 13147–13155.
7. Dong, X., et al. A novel rutile TiO_2/$AlPO_4$ core–shell pigment with substantially suppressed photoactivity and enhanced dispersion stability. *Powder Technology*, 2020, **366**: pp. 537–545.
8. Yao, B., et al. Synthesis, characterization, and optical properties of near-infrared reflecting composite inorganic pigments composed of TiO_2/CuO core–shell particles. *Australian Journal of Chemistry*, 2018, **71**(5): pp. 373–379.
9. Huseien, G.F., K.W. Shah, and A.R.M. Sam. Sustainability of nanomaterials based self-healing concrete: An all-inclusive insight. *Journal of Building Engineering*, 2019, **23**: pp. 155–171.
10. Izu, N., et al. Decreasing the shell ratio of core-shell type nanoparticles with a ceria core and polymer shell by acid treatment. *Solid State Sciences*, 2018, **85**: pp. 32–37.
11. Ahmed, N., et al. Evaluation of new core-shell pigments on the anticorrosive performance of coated reinforced concrete steel. *Progress in Organic Coatings*, 2020, **140**: p. 105530.

12. Sanchis-Gual, R., et al. Plasmon-assisted spin transition in gold nanostar@ spin cross-over heterostructures. *Journal of Materials Chemistry C*, 2021, **9**(33): pp. 10811–10818.

13. Torres-Cavanillas, R., et al. Design of bistable gold@ spin-crossover core–shell nanoparticles showing large electrical responses for the spin switching. *Advanced Materials*, 2019, **31**(27): p. 1900039.

14. Ahmed, N.M., M.G. Mohamed, and W.M. Abd El-Gawad. Corrosion protection performance of silica fume waste-phosphates core-shell pigments. *Pigment & Resin Technology*, 2018, **47**(3); 1–14.

15. Sertchook, H. and D. Avnir. Submicron silica/polystyrene composite particles prepared by a one-step sol– gel process. *Chemistry of Materials*, 2003, **15**(8): pp. 1690–1694.

16. Zhuang, Z., W. Sheng, and Y. Yan. Synthesis of monodispere Au@ Co_3O_4 core-shell nanocrystals and their enhanced catalytic activity for oxygen evolution reaction. *Advanced Materials*, 2014, **26**(23): pp. 3950–3955.

17. Zhong, Z., et al. Preparation of mesoscale hollow spheres of TiO_2 and SnO_2 by templating against crystalline arrays of polystyrene beads. *Advanced Materials*, 2000, **12**(3): pp. 206–209.

18. Samal, A.K., et al. Size tunable Au@ Ag core–shell nanoparticles: Synthesis and surface-enhanced raman scattering properties. *Langmuir*, 2013, **29**(48): pp. 15076–15082.

19. Parashar, M., V.K. Shukla, and R. Singh. Metal oxides nanoparticles via sol–gel method: A review on synthesis, characterization and applications. *Journal of Materials Science: Materials in Electronics*, 2020, **31**(5): pp. 3729–3749.

20. Dugay, J., et al. Charge mobility and dynamics in spin-crossover nanoparticles studied by time-resolved microwave conductivity. *The Journal of Physical Chemistry Letters*, 2018, **9**(19): pp. 5672–5678.

21. Ahmed, J., et al. Microemulsion-mediated synthesis of cobalt (pure fcc and hexagonal phases) and cobalt–nickel alloy nanoparticles. *Journal of Colloid and Interface Science*, 2009, **336**(2): pp. 814–819.

22. El-Safty, S.A.. Synthesis, characterization and catalytic activity of highly ordered hexagonal and cubic composite monoliths. *Journal of Colloid and Interface Science*, 2008, **319**(2): pp. 477–488.

23. Ghosh Chaudhuri, R. and S. Paria. Core/shell nanoparticles: Classes, properties, synthesis mechanisms, characterization, and applications. *Chemical Reviews*, 2012, **112** (4): pp. 2373–2433.

24. Han, W., et al. Synthesis and shape-tailoring of copper sulfide/indium sulfide-based nanocrystals. *Journal of the American Chemical Society*, 2008, **130**(39): pp. 13152–13161.

25. Lee, H., et al. Shape-controlled nanocrystals for catalytic applications. *Catalysis Surveys from Asia*, 2012, **16**(1): pp. 14–27.

26. Libor, Z. and Q. Zhang. The synthesis of nickel nanoparticles with controlled morphology and SiO_2/Ni core-shell structures. *Materials Chemistry and Physics*, 2009, **114** (2–3): pp. 902–907.

27. Kim, Y., et al. Silica effect on coloration of hematite nanoparticles for red pigments. *Chemistry Letters*, 2009, **38**(8): pp. 842–843.

28. Liu, R. and R.D. Priestley. Rational design and fabrication of core–shell nanoparticles through a one-step/pot strategy. *Journal of Materials Chemistry A*, 2016, **4**(18): pp. 6680–6692.

29. Chen, S., et al. Preparation and characterization of monodispersed spherical Fe_2O_3@ SiO_2 reddish pigments with core–shell structure. *Journal of Advanced Ceramics*, 2019, **8**(1): pp. 39–46.

30. Dodd, A.C.. A comparison of mechanochemical methods for the synthesis of nanoparticulate nickel oxide. *Powder Technology*, 2009, **196**(1): pp. 30–35.

31. Deng, W., et al. Formation of ultra-fine grained materials by machining and the characteristics of the deformation fields. *Journal of Materials Processing Technology*, 2009, **209**(9): pp. 4521–4526.
32. Salari, M., P. Marashi, and M. Rezaee. Synthesis of TiO_2 nanoparticles via a novel mechanochemical method. *Journal of Alloys and Compounds*, 2009, **469**(1–2): pp. 386–390.
33. Sasikumar, R. and R. Arunachalam. Synthesis of nanostructured aluminium matrix composite (AMC) through machining. *Materials Letters*, 2009, **63**(28): pp. 2426–2428.
34. Wang, Y., K. Cai, and X. Yao. Facile synthesis of PbTe nanoparticles and thin films in alkaline aqueous solution at room temperature. *Journal of Solid State Chemistry*, 2009, **182**(12): pp. 3383–3386.
35. Yoo, S.-H., L. Liu, and S. Park. Nanoparticle films as a conducting layer for anodic aluminum oxide template-assisted nanorod synthesis. *Journal of Colloid and Interface Science*, 2009, **339**(1): pp. 183–186.
36. Oldenburg, S., et al. Nanoengineering of optical resonances. *Chemical Physics Letters*, 1998, **288**(2–4): pp. 243–247.
37. Daniel, M.-C. and D. Astruc. Gold nanoparticles: Assembly, supramolecular chemistry, quantum-size-related properties, and applications toward biology, catalysis, and nanotechnology. *Chemical Reviews*, 2004, **104**(1): pp. 293–346.
38. Patra, J.K. and K.-H. Baek. Green nanobiotechnology: Factors affecting synthesis and characterization techniques. *Journal of Nanomaterials*, 2014, **14**, 1–12.
39. Hayes, R., et al. Core–shell particles: Preparation, fundamentals and applications in high performance liquid chromatography. *Journal of Chromatography A*, 2014, **1357**: pp. 36–52.
40. Zhang, Y., et al. Learning from ancient Maya: Preparation of stable palygorskite/methylene blue@ SiO_2 Maya Blue-like pigment. *Microporous and Mesoporous Materials*, 2015, **211**: pp. 124–133.
41. Mao, W.-X., et al. Core–shell structured Ce_2S_3@ ZnO and its potential as a pigment. *Journal of Materials Chemistry A*, 2015, **3**(5): pp. 2176–2180.
42. Liu, J., et al. Yolk/shell nanoparticles: New platforms for nanoreactors, drug delivery and lithium-ion batteries. *Chemical Communications*, 2011, **47**(47): pp. 12578–12591.
43. Niu, H.-Y., et al. A core–shell magnetic mesoporous silica sorbent for organic targets with high extraction performance and anti-interference ability. *Chemical Communications*, 2011, **47**(15): pp. 4454–4456.
44. Insin, N., et al. Incorporation of iron oxide nanoparticles and quantum dots into silica microspheres. *ACS Nano*, 2008, **2**(2): pp. 197–202.
45. Wang, J., et al. Accurate control of multishelled Co_3O_4 hollow microspheres as high-performance anode materials in lithium-ion batteries. *Angewandte Chemie*, 2013, **125**(25): pp. 6545–6548.
46. Lai, X., et al. General synthesis and gas-sensing properties of multiple-shell metal oxide hollow microspheres. *Angewandte Chemie International Edition*, 2011, **50**(12): pp. 2738–2741.
47. Wang, D., et al. Structurally ordered intermetallic platinum–cobalt core–shell nanoparticles with enhanced activity and stability as oxygen reduction electrocatalysts. *Nature Materials*, 2013, **12**(1): pp. 81–87.
48. Chae, H.S., et al. Core-shell structured Fe_3O_4@ SiO_2 nanoparticles fabricated by sol–gel method and their magnetorheology. *Colloid and Polymer Science*, 2016, **294**(4): pp. 647–655.
49. Guo, X., et al. Sol–gel emulsion synthesis of biphotonic core–shell nanoparticles based on lanthanide doped organic–inorganic hybrid materials. *Journal of Materials Chemistry*, 2012, **22**(13): pp. 6117–6122.

50. Li, Y.-M., et al. Preparation and thermal stability of silica layer multicoated γ-Ce$_2$S$_3$ red pigment microparticles. *Surface and Coatings Technology*, 2018, **345**: pp. 70–75.

51. Nikolaev, P., et al. Gas-phase catalytic growth of single-walled carbon nanotubes from carbon monoxide. *Chemical Physics Letters*, 1999, **313**(1–2): pp. 91–97.

52. El-Gendy, A., et al. The synthesis of carbon coated Fe, Co and Ni nanoparticles and an examination of their magnetic properties. *Carbon*, 2009, **47**(12): pp. 2821–2828.

53. Mendes, R.G., et al. Synthesis and toxicity characterization of carbon coated iron oxide nanoparticles with highly defined size distributions. *Biochimica et Biophysica Acta (BBA)-General Subjects*, 2014, **1840**(1): pp. 160–169.

54. Stöber, W., A. Fink, and E. Bohn. Controlled growth of monodisperse silica spheres in the micron size range. *Journal of Colloid and Interface Science*, 1968, **26**(1): pp. 62–69.

55. Bogush, G., M. Tracy, and C. Zukoski IV. Preparation of monodisperse silica particles: Control of size and mass fraction. *Journal of Non-Crystalline Solids*, 1988, **104**(1): pp. 95–106.

56. Drašar, P., L. David, Z. Marcos. The sol-gel handbook: Synthesis, characterization and applications. *Chemické Listy*, 2016, **110**(3): pp. 229–230.

57. Du, G.-H., et al. Characterization and application of Fe$_3$O$_4$/SiO$_2$ nanocomposites. *Journal of Sol-Gel Science and Technology*, 2006, **39**(3): pp. 285–291.

58. Li, Z., et al. Synthesis of Zn$_2$SiO$_4$@ ZnO core-shell nanoparticles and the effect of shell thickness on band-gap transition. *Materials Chemistry and Physics*, 2020, **240**: p. 122144.

59. Tsuji, M., et al. Epitaxial growth of Au@ Ni core–shell nanocrystals prepared using a two-step reduction method. *Crystal Growth & Design*, 2011, **11**(5): pp. 1995–2005.

60. Fan, F.-R., et al. Epitaxial growth of heterogeneous metal nanocrystals: From gold nano-octahedra to palladium and silver nanocubes. *Journal of the American Chemical Society*, 2008, **130**(22): pp. 6949–6951.

61. Tsuji, M., et al. Synthesis of Ag@ Cu core–shell nanoparticles in high yield using a polyol method. *Chemistry Letters*, 2010, **39**(4): pp. 334–336.

62. Sharma, R. and S. Tiwari. Synthesis of fly ash based core-shell composites for use as functional pigment in paints. In: *AIP Conference Proceedings*, 2016. AIP Publishing LLC, USA, 1724(1), p. 020083.

63. Zhang, J., et al. Synthesis of core–shell acrylic–polyurethane hybrid latex as binder of aqueous pigment inks for digital inkjet printing. *Progress in Natural Science: Materials International*, 2012, **22**(1): pp. 71–78.

64. Galogahi, F.M., et al. Core-shell microparticles: Generation approaches and applications. *Journal of Science: Advanced Materials and Devices*, 2020, **5**(4): pp. 417–435.

65. Cao, L., et al. Inorganic–organic hybrid pigment fabricated in the preparation process of organic pigment: Preparation and characterization. *Dyes and Pigments*, 2015, **119**: pp. 75–83.

66. Bongur, R., et al. Red 33 dye co-encapsulated with cetyltrimethylammonium in mesoporous silica materials. *Dyes and Pigments*, 2016, **127**: pp. 1–8.

67. He, X., et al. Fabrication of highly dispersed NiTiO$_3$@TiO$_2$ yellow pigments with enhanced NIR reflectance. *Materials Letters*, 2017, **208**: pp. 82–85.

68. Wang, F., et al. Structural coloration pigments based on carbon modified ZnS@ SiO$_2$ nanospheres with low-angle dependence, high color saturation, and enhanced stability. *ACS Applied Materials & Interfaces*, 2016, **8**(7): pp. 5009–5016.

69. Guan, L., et al. Facile preparation of highly cost-effective BaSO$_4$@ BiVO$_4$ core-shell structured brilliant yellow pigment. *Dyes and Pigments*, 2016, **128**: pp. 49–53.

70. Sultan, S., K. Kareem, and L. He. Synthesis, characterization and resistant performance of α-Fe$_2$O$_3$@SiO$_2$ composite as pigment protective coatings. *Surface and Coatings Technology*, 2016, **300**: pp. 42–49.

71. Zou, J. and W. Zheng. TiO_2@ $CoTiO_3$ complex green pigments with low cobalt content and tunable color properties. *Ceramics International*, 2016, **42**(7): pp. 8198–8205.

72. Liu, S.-G., et al. Enhanced high temperature oxidization resistance of silica coated γ-Ce_2S_3 red pigments. *Applied Surface Science*, 2016, **387**: pp. 1147–1153.

73. Sadeghi-Niaraki, S., et al. Nanostructured Fe_2O_3@TiO_2 pigments with improved NIR reflectance and photocatalytic ability. *Materials Chemistry and Physics*, 2019, **235**: p. 121769.

74. Li, Y., et al. Preparation and characterization of the Sr^{2+}-doped γ-Ce_2S_3@c-SiO_2 red pigments exhibiting improved temperature and acid stability. *Applied Surface Science*, 2020, **508**: p. 145266.

75. Li, Y., et al. Synthesis and characterization of aluminum-based γ-Ce_2S_3 composite red pigments by microemulsion method. *Journal of Alloys and Compounds*, 2020, **812**: p. 152100.

76. Jang, H.-S., H.-S. Kang, and S.-Y. So. Color expression characteristics and physical properties of colored mortar using ground granulated blast furnace slag and White Portland Cement. *KSCE Journal of Civil Engineering*, 2014, **18**(4): pp. 1125–1132.

77. Ahmed, N.M., H.T.M. Abdel-Fatah, and E.A. Youssef. Corrosion studies on tailored ZnCo aluminate/kaolin core–shell pigments in alkyd based paints. *Progress in Organic Coatings*, 2012, **73**(1): pp. 76–87.

2 Modification of Cement-Based Materials with Nanoparticles

2.1 INTRODUCTION

With the exception of water, concrete in various forms remains the most commonly consumed primary manufactured material globally [1–3]. Presently, it is estimated that 4.4 billion tonnes of concrete are produced worldwide annually, with this figure expected to surpass 5.5 billion tonnes by 2050, which is attributable to the increasing urbanisation of developing nations [4]. According to one survey, it is likely that by 2050, there will be a 23% increase from 2014 in the demand for cement-based concretes [5]. This is due to the ongoing and rapid developments in industrialised countries including China and India and regions such as Northern Africa, Southeast Asia and the Middle East. Huge quantities of cement-based concretes are necessary to meet the demand of the new infrastructures being established in developing countries. These include the new highways, bridges, businesses, homes, colleges, schools and hospitals that are being constructed in urban areas as a result of the recent economic growth [1,6].

One type of cement is ordinary Portland cement (OPC). It is manufactured in a rotary kiln by heating limestone or chalk with clay to a temperature of 1450°C, which creates solid clinker knots that are subsequently grinded with some gypsum via grinding balls, thereby producing conventional cement. There are numerous significant environmental concerns linked with the manufacture of OPC including the vast amount of energy consumed in the heating process, the swift destruction of the landscape, the high level of dust produced during transportation, the noise pollution generated in the quarries and the production of raw materials [7]. Many studies have shown that cement production is one of the key manufacturing industries to which a substantial proportion of greenhouse gas emissions can be attributed. Overall, it is the third-highest energy-consuming industrial field and the second-highest industrial CO_2 (carbon dioxide) emitter, which accounts for 7% of all energy consumption and almost 7% of all CO_2 emissions [8]. Of late, CO_2 emissions have caused a new problem for the cement industry: per 1 tonne of cement production, there is circa 1 tonne of CO_2 emissions. At the same developmental pace, these CO_2 emissions are predicted to increase by 13.5% (2.596 Mt) in 2030 and 5.7% (2.416 Mt) in 2050 over the 2010 levels [9]. Furthermore, during the cement production process, huge amounts of greenhouse gases such as CO_2, SO_x and NO_x are released [10,11], and there is also a substantial amount of limestone usage and huge energy consumption. Additionally, the manufacture of cement has a seriously detrimental effect on environmental

DOI: 10.1201/9781003281504-2

sustainability due to the huge amounts of dust it releases into the atmosphere, which is highly damaging to groundwater, crop production and living systems.

Advancements made recently in preparation and characterisation approaches linked to nanoscience and nanotechnology have enabled the reconstruction of a range of functional man-made materials at low dimensions through maintaining their fundamental characteristics [12]. There are two methods of synthesising nanoscale materials: (i) top-down [13] and (ii) bottom-up [14]. A variety of factors dictate which method is selected, such as efficacy of emergent properties, cost-effectiveness, suitability and eco-friendliness [13]. For example, milling is a flexible top-down method utilised to create tiny nanoscale structures with emerging properties, while avoiding the need to alter the atomic-scale characteristics. In contrast, bottom-up approaches enable the matter to be designed and controlled at an atomic level according to the chemical reaction routes [12–14]. Top-down technique is more cost-effective and less complex. However, bottom-up approach can produce high standard nanomaterials from their bulk counterparts in an immaculate manner although more expensive, requiring complex procedures [12]. Nanoparticles (NPs) can exist in different dimensions, shapes and sizes, such as cubes, cylinders, discs, rings, spheres and triangles.

Figure 2.1 depicts the NPs' classification scheme according to their type of origin, such as dimensions and nanostructures (nanocrystalline, polymer or non-polymer) [15]. The NPs can be generally categorised in two ways according to their composition: (i) organic carbon compounds (e.g., polymeric) and (ii) inorganic compounds (e.g., metal oxides). Organic NPs, which include dendrimer, ferritins and hollow spheres, are usually biodegradable and non-toxic (e.g., micelle and liposome). Conversely, inorganic NPs are mainly comprised of metals (e.g., aluminium (Al), gold (Au), cadmium (Cd), cobalt (Co), copper (Cu), iron (Fe), silicon (Si) and zinc (Ze)) and metal oxides (SiO_2, Fe_3O_4, ZnO_2, FeO_2, TiO_2 and CeO_2). Regardless of origin, all nanomaterials demonstrate unique properties not observed in their bulk counterparts [16,17]. Levels of CO_2 emissions and energy consumption can be decreased by employing a nanotechnology approach and reusing the industrial waste or by-products (e.g., iron blast furnace slag, fly ash and nano-particulates). This facilitates meeting the construction industries' sustainability standards in relation to the environment, cost and society (see Figure 2.2).

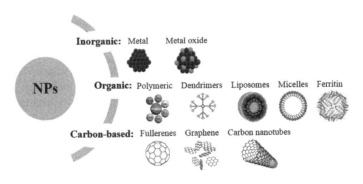

FIGURE 2.1   Nanoparticles classification based on their chemical composition [16].

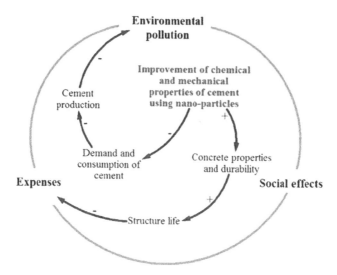

FIGURE 2.2 Causal loop diagram indicating the nanoparticles' role in sustainable development of the construction industry [18].

With consideration of the huge significance of diverse nanomaterials in sustainable cement-based concretes, the purpose of this chapter is to offer an in-depth picture of the potential of the application of nanotechnology and nanomaterials in the cement industry. Furthermore, it aims to deliver a clear comprehension of how the use of industrial waste/by-product nanomaterials as additives can enhance the cement-based concretes' microstructures, resulting in sustainable advancements in the global construction industries. In addition, this chapter describes the microscopic mechanisms of the impact of these nano-systems on the mechanical traits of the altered concretes in relation to their workability, durability, setting time, strength traits and durability indices.

2.2 NANOTECHNOLOGY AND NANOMATERIALS

Figure 2.3 presents the different nanoscale materials and their various uses in advanced technologies. Nanotechnology could be a major innovation in construction, leading to a deeper comprehension of the workings of building materials [19]. The outstanding scientific breakthroughs in the nanoscience and nanotechnology fields (both microscopy instruments and materials) have enabled the optimisation of nanoscale materials' distinctive characteristics to improve the basic properties of traditional materials such as concrete, glass, metals, paints, plastics and wood, and to utilise waste materials including coal product, waste ceramic and glass, and rice husk ash. Currently, nanomaterials are utilised in a variety of diverse industries and applications; for instance, the manufacturer of ceramic, glass, and steel, medicine, water treatment, and the coating and isolating of roofs and windows [20,21]. The significant progress made in nanotechnology and nanomaterials has facilitated the

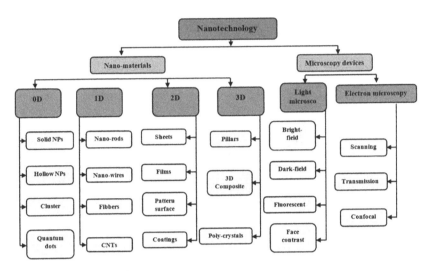

FIGURE 2.3 Various nanoscale materials and their diverse applications in advanced technologies [3].

integration of innovative properties into the basic material structure. Consequently, fresh new concepts to address concrete durability issues are likely to emerge, and future constructions can potentially be completed with substantially lower service and maintenance expenses [22].

Studies have shown that different nanotechnology-based products have varied and unique characteristics that can majorly alter the current construction sectors' practices, resulting in significantly enhanced planning and design concepts [19]. For example, nanostructures of Al_2O_3, Ag, Cu, SiO_2 and TiO_2 can be utilised in a variety of ways such as coating various construction components (e.g., floor, roof and toilet) to make them waterproof, to decrease corrosion and to protect against UV radiation [22]. Different studies of green building technology have reported that superior antimicrobial properties can be achieved when using paints that contain Ag and Cu nanoparticles [23,24]. Furthermore, enhanced electrical, mechanical and thermal properties were observed in plastics that had been reinforced with carbon nanotube (CNT) [25]. In addition, CNT with isodal or cylindrical nanostructure molecules (with ratios of length to diameter reaching 132,000,000:1) demonstrates remarkable features that are highly beneficial to a broad range of applications, such as electronic devices, nanotechnology, optical devices and material science-based technologies [26].

When coated in the glass windowpanes (silica aerogel), nanomaterials demonstrate thermal insulation and self-cleaning properties, thereby minimising the flow of heat in the hot atmosphere and controlling the transfer of heat through the window glass [27]. By applying advanced nanotechnology to make basic changes to the properties of the materials, nanoclay can be widely utilised in the construction fields at a significantly lower cost, thereby driving the cost-effectiveness and sustainability of the construction. Montmorillonite clay is a viable replacement for cement due to its distinctive structure and high quantity of silicates [26]. Moreover, resistance to wear and strength can be increased by using nanostructured-based ceramic and steel composites. Furthermore, the addition of nanoparticles to concrete can enhance its properties and overall performance. Most research of nanostructured materials in concretes concurs that early mechanical strength and general bulk properties are notably improved.

2.3 NANOMATERIALS-MODIFIED CEMENT BINDER

Integrating the nanomaterials into the cement-based concretes creates sustainable building materials that can improve bulk properties and microstructures [28]. It is widely acknowledged that the particular nature of the nanomaterials' surface area-to-volume ratio enables them to stimulate and accelerate the rate of pozzolanic chemical reaction in the concrete mixture. The pozzolanic property (lime reacts in water) is present in many nanoparticles such as clay and silica. As shown in Figure 2.4, the molecules' amorphous structure means that it is capable of binding the loose materials together. As a result, it speeds up the hydration reaction and enhances the early mechanical strengths. Furthermore, the morphological properties are also enhanced which is attributable to the interfacial connectivity between different aggregates and cement binder particles. Essentially, the nanostructured elements can function as super-filler, which results in lower porosity in the concrete and greater packing density. Basically, the nanomaterials-modified cement binders' enhanced morphologies help to increase the durability and strength of the concrete [29]. Most research in the area of nanomaterials-modified cement binders and concretes references the widespread employment of nanosilica (NS) [30–33].

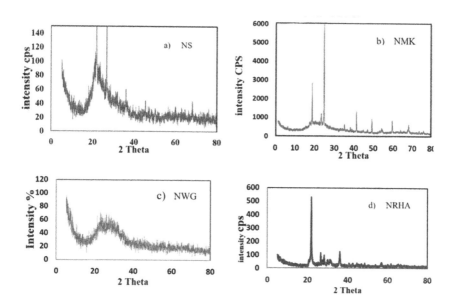

FIGURE 2.4 XRD patterns of (a) NS, (b) nanometakoline (NMK), (c) nano waste glass (NWG) and (d) nanorice husk ash (NRHA) [34].

Research outcomes showed that the cement hydration reaction rate and the microstructure improved, subsequently generating superior bulk density, greater early strength, increased compactness and advanced durability indices. In addition, it is beneficial in restricting hydroxide calcium $(Ca(HO)_2)$ leaching. This is significant, as this leaching is recognised as being a major factor in concrete deterioration, and therefore, restricting it leads to a longer service life. SiO_2 nanoparticles have several functions including as the filler to improve the microstructures and interfacial interactions and contributing to the pozzolanic reactions that greatly affect the performance of cement-based concretes. There have been numerous and comprehensive studies of the different characteristics of the nano-Al_2O_3 [35–41], nano-TiO_2 [42–46], CNTs [47,48] and nanoclay [49]. The outcomes have proven significantly enhanced mechanical strength and durability in severe environments. In relation to the microstructural traits of the nanomaterials (included cement binders), the improvements were deemed to be due to the cement hydration products and aggregates' improved microstructures and bonds reinforcement [50]. Concretes with nano-Al_2O_3 displayed comparable strength and durability to that of concretes with NS. Furthermore, cement binders with nanoTiO$_2$ and CNTs presented fresh photocatalytic and piezoelectric characteristics that are beneficial in a range of applications [51–53]. The optimum nanomaterials for improving cement-based concretes' mechanical characteristics while simultaneously displaying innovative traits (e.g., electromagnetic shielding, self-sensing) are widely considered to be CNTs and carbon nanofibres (CNFs) [48,54].

Table 2.1 presents the impact of the integration of various nanomaterials on cement-based concretes' innovative properties, in which the hydration mixtures' heat was seen to substantially improve when NPs were present. As shown in Figure 2.5,

TABLE 2.1
Performance of Various Nanomaterials-Activated Cement Binders

Refs.	Type of NPs	Level, %	Findings	
			Slump Flow (Reduction, %)	Setting Time (Reduction, %)
[60]	SiO_2	3	58	-
[61]	TiO_2	5	-	30–45
[62]	ZnO	0.2	4	-
[63]	ZnO	5	-	17–22
[64]	TiO_2	5	31.2	-
[65]	SiO_2	4	57.2	12.5
[66]	Clay	3	14.3	-
[67]	Al_2O_3	2	70	40–55
[68,69]	Fe_2O_3	2	60	40–55
[70]	SiO_2	3	35–37.5	-
[41]	SiO_2	3	-	60–55
[41]	Al_2O_3	3	-	7
[55]	SiO_2	2	-	22–28

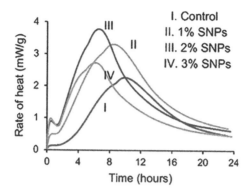

FIGURE 2.5 Silica NPs (SNPs) percentage-dependent hydration rate of various concrete [60].

this was due to the NPs' broader surface area creating a raised hydration peak temperature. It signified the substantial effect the nanomaterials have on the rate of acceleration reaction, resulting in lower setting times [55,56]. Land and Stephan maintain that the increased hydration heat is thought to be chiefly attributable to the increased surface area [57,58]. Numerous studies have reported that the hydration heat rises as the quantity of nanoparticles increases, thereby necessitating greater water adsorption. Research by Mukharjee and Barai has shown that their fineness has a massive impact on the setting times of different nanomaterials inserted cement binders [59]. It has also been found that increasing the fineness of the nano-cementitious materials can enhance the cement pastes' chemical reactivity, resulting in greater durability and strength. The availability of numerous nucleation centres offered by the NPs is the main reason for the observed chemical reaction rate, in which the rate of hydration reaction is increased, the slump flow is improved and the setting time is lowered.

Reports state that setting time is reduced by increasing the nano-SiO_2 content in the paste, which is because of the profusion of free Ca ions released from the cement or slag that produce extra C-S-H [41,71]. Some reports maintain that there is a correlation between increased contents of nanomaterials and decreased workability of the final product [72–75]. In comparison to the pure cement paste, those incorporated with the NPs of SiO_2, SF, Al_2O_3, TiO_2 and C demonstrated substantially lower flowability [76,77]. The reason for this is that NPs are finer, thereby facilitating greater cohesiveness in the concretes or mortars. A study conducted by Hosseini et al. found that the addition of NS nanoparticles produced a notable decrease in workability [78]. The addition of 1.5% and 3% colloidal NS content to 100% recycled coarse aggregate modified concrete causes the slump flow to decrease by 47.1% and 70.59%, respectively. The high reactivity stimulated from the silica NPs' wider specific surface area is attributed to the concreates' observed decrease in workability.

Research conducted by Jalal et al. [79,80] identified lower flowability in the concrete when SF and NS are present. In addition, several concrete properties (such as the consistency of the mix and its bleeding and segregation) were found to have improved. According to Joshaghani et al. [81], all TiO_2, Al_2O_3 and Fe_2O_3 nanoparticles created a notable decrease in slump flow. An important point here is that as the percentage of nanomaterials increases, the percentage of high-range water decreasing admixture (HRWRA) also increases, thereby accomplishing a slump flow of 650 ± 25 mm until it achieved a height of 36% for 5% TiO_2 [82]. Most researchers agreed that the reduced flowability was because of the nanoparticles' high reactivity, which have a larger specific surface area than cement; thus, more water is absorbed, decreasing the concretes' slump [83].

Over the last 20 years, nanomaterials have become more prevalent in cementitious systems which is because of their capacity to enhance mechanical strength. Much research has been carried out to examine how diverse nanomaterials improve the strength indexes of a nanostructures-modified concrete system [47,61,83–91]. Table 2.2 depicts a rise in the early ages compressive (CS), flexural (FS) and splitting tensile strength (STS) of the cement mortars or pastes integrated with diverse nano-oxide materials including alumina, silica, silica fume (SF), Fe_2O_3/Fe_3O_4, zinc, Zr, $CaCO_3$, Cu_2O_3/CuO, CNTs and clay. Research conducted by Zhang et al. [92] examined the impact on white cements' (w/c = 0.5) hydration and strength properties when replacing 1% of NS (15 and 50 nm), nano-Fe_2O_3 (50 nm) and nano-NiO (15 nm), respectively. The researchers determined that following 3 days of curing, there were increases on the control sample of 23.9%, 22.8%, 21.9% and 20.5% in the compressive strength of the four samples (50 nm Fe_2O_3, 15 nm NiO, 15 nm SiO_2 and 50 nm SiO_2), respectively. After 28 days, the compressive strength increases were recorded as 10.2%, 9.5%, 10.5% and 11.2%, respectively. This was attributed to the enhanced pozzolanic reactivity, filling impact, extra nucleation sites created by NPs, least Ca(-OH)$_2$ content and decreasing volume of pores.

Comparable outcomes were achieved in a study by Said et al. [56], who utilised a ternary mix of 30% fly ash cement replacement and NS. They found that the addition of 3%–6% NS caused an average compressive strength increase over the control samples for all groups, with results of 18%, 14% and 36% after 3, 7 and 28 days of curing, respectively. The improved strength properties were attributed to the pozzolanic reaction and NS filling behaviour. Kaur et al. [105] investigated the effect of nano-metakaolin on fly ash-based mortars' strength performance by replacing 0%, 2%, 4%, 6%, 8% and 10% nano-metakaolin with fly ash. Following 3 days of ambient curing, the 4% replacement sample obtained circa 70%–80% of the compressive strength after 28 days curing. When compared to the control group, the increases in compressive strength after 3, 7, 14 and 28 days of curing were reported as 26.5%, 21.4%, 21.4% and 22.7%, respectively. The effect of increasing the nano-metakaolin content to 10% was a decrease in the compressive strength of around 1%–2% compared to the control sample for all curing durations. Another study by Nuaklong et al. [106] found that substituting 30% of nano-metakaolin achieved maximum compressive strength increase in high-calcium fly ash-based mortar. As per the researchers, this variance was due to the disparity in the CaO content and the alkaline activator solution ratio of the mixtures.

Gunasekara et al. [107] examined the effect of incorporating NS into high-volume fly ash (HVFA)-blended concretes. They found that adding 3% NS into the mixture with FA (65% and 80% cement) resulted in increases in the compressive strength after 7 days curing of 50% and 98.6%, respectively, over the control sample, and 10.3% and 35.9%, respectively, after 28 days curing. Hou et al. [108,109] produced

TABLE 2.2
Effects of Various Nanomaterials on the Strength Performance of Concretes

Type of NPs	Replacement, %	Rate of Increment, %			Refs.
		CS, %	FS, %	STS, %	
SiO_2	5	45@3days	31.4@3days	-	[87]
		29.7@7days	27.4@7days		
		10.6@28days	9.8@28days		
SiO_2	6	33.5@28days	-	56@28days	[93]
Fe_2O_3	3	26@28days	18@28days	-	[94]
TiO_2	3	15@28days	18@28days	20@28days	[95]
CNT	1.2	11.9@28days	17.3@28days	15.2@28days	[96]
Fe_2O_3	1	15.5@28days	18.2@28days	55.6@28days	[68,69]
Fe_2O_3	1–5	No significant increase	-	-	[97]
SiO_2-Al_2O_3	5	17@28days	14@28days	-	[92]
Al_2O_3	1	10@28days	16@28days	22@28days	[77]
Fe_2O_3	5	57@28days	46@28days	75@28days	[98]
SiO_2	6	32@28days	-	20@28days	[99]
TiO_2-graphine nanosheets	0.03	28.6@7days	26.6@28days	-	[100]
		11.3@28days			
CNTs	0.5	32@28days	28@28days	-	[101]
TiO_2	1	18@28days	10.3@28days	-	[102]
SiO_2	4	58@3days	-	-	[97]
		22@90days			
CuO	3	51@3days	-	-	[97]
		21@90days			
Al_2O_3	3	16.6@28days	16.7@28days	-	[103]
		18.7@90days	18.8@90days		
SiO_2	4	18@28days	40@28days	35@28days	[50]
Combined SF and SiO_2	10% SF + 2% SiO_2	30@7days	58@90days	33@90days	[79]
		73@90days			
Al_2O_3	1.5	55@28days	-	26@28days	[104]
ZrO_2	1.5	20@28days	-	11@28days	[104]
TiO_2	1.5	23@28days	-	17@28days	[104]
Fe_3O_4	1.5	29@28days	-	27@28days	[104]
NGP	5	17@28days	14@28days	22@28days	[91]

similar findings in their investigation of mortar. They attributed the compressive strength increase primarily to the presence of fine NS particles that function as seeds offering extra nucleation sites which accelerate the mixtures' hydration process. Furthermore, the researchers also assessed the NSs' pozzolanic trait and found that it reacted with the lime to generate additional calcium-silicate hydrate (C-S-H) gel. This occurred for two main reasons. Firstly, following the dissolution of the NS in water, H_2SiO_4 was formed, which subsequently reacted with the Ca^{2+}, thereby producing C-S-H gel. Secondly, as mentioned, the presence of NS functioned as seeds inside pores offering extra nucleation sites, thereby improving the pozzolanic characteristics [110]. When cement was substituted with high-content nanoparticle-based waste materials (SiO_2-Al_2O_3) as a green concrete/mortar, it led to a positive outcome. Table 2.3 presents additional results from more recent studies examining the impact of various nanomaterials on the compressive and flexural strength development of concretes following 28 days of curing.

Figure 2.6 illustrates the SEM images of the pure OPC, the OPC with 3% of NS and the OPC with 3% of NT, respectively. The micrographs were comprised of different sized CH crystals for the specimens incorporating nano-TiO_2 (3 μm) and NS (5 μm) particles which were smaller in size than those contained in the pure OPC sample. These outcomes verified that NPs are pivotal in relation to filling the large pores and creating additional nucleation sites for CH consumption, controlling the CH crystals' orientation development and decreasing the total crystal size [124]. Controlling the CH crystals' (hexagonal-platelet morphology) development decreases the likelihood of cracks appearing at the interfacial transition zone (ITZ),

TABLE 2.3

CS vs. FS Development of Concrete Specimens Containing Different Types of Nanomaterials

Refs.	Materials	%	w/c	Increment on CS, % per Day		Increment on FS, % per Day	
				7	28	7	28
[111]	NS	2	0.40	17	36	-	-
	NS	3	0.40	11	16		
[112]	NS	3	0.45	-	50	-	74.6
[113]	NS	3	0.50	82.5	48	-	-
		2	0.50	43.5	58.5	-	-
[114]	NS	3	0.45	-	43.4	-	20.3
[115]	NS	2	0.40	22	24	12	16
		4	0.40	26	31	18	25
[116]	Colloidal	10	0.34	-	48.1	-	-
	NS	7.5	0.35	-	45.8	-	-
[117]	MK	10	0.40	-	16.5	-	28.4
	MK+CNT	10+0.05	0.40	-	12.2	-	28.6
[118]	NS	3	0.50	-	18.8	-	-
	NA	3		-	10	-	-
	NS+NA	6		-	18.5	-	-
[119]	TiO_2	3	0.58	-	-	-	61.9
[120]	TiO_2	2	0.48	-	52.4	-	-
[121]	$CaCO_3$	3.2	0.18	-	10	-	20
	NS	1	0.18	-	6	-	12
[122]	NA	1	0.50	10.7	15.5	6452	41.1
[123]	Fe_2O_3	3	0.50	-	67.3	-	58.3

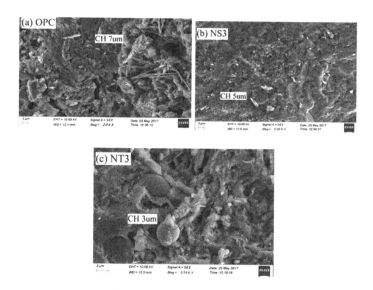

FIGURE 2.6 SEM images of (a) OPC, (b) OPC with 3% of NS and (c) OPC with 3% of NT [124].

which is recognised as being the zone that manages mechanical properties failure (weak zones). The inclusion of NPs decelerated the $Ca(OH)_2$ crystals' rate of development. This was due to the NPs' intense reactivity and wide surface area in the concrete matrix. It was established that the incorporation of diverse NPs in the OPC matrix enabled the deceleration of the rate of development of CH crystals in the simulated transition regions, which manage the generation of extra C-S-H gel and facilitate gaps/slash pores filling agents in the OPC matrix [82].

Generally speaking, there are four key effects of nanomaterials in the modified concretes/mortars. Firstly, nanomaterials function as pore fillers in the concrete matrix, thereby increasing its compactness. Secondly, nanoparticles have a high electrostatic force which stimulates hydration in the cement mix and creates additional nucleation sites which subsequently produce additional C-S-H gel clusters. Thirdly, microcracks and filling pores are elements of nanoparticles that enhance homogeneity and make the network more compact than conventional OPC-based concretes that do not incorporate NPs. Fourthly, NPs' high reactivity means that they have a chemical reaction with $Ca(OH)_2$ and consume more of them. Hence, they are amassed in the micropores and ITZ, thereby prompting a refinement process in the microcracks in the initial stages which enhances the concrete matrix' microstructures.

2.4 SUSTAINABILITY PERFORMANCE

Chloride exposure, which can cause embedded steel to corrode, is one trigger of long-term corrosion in major concrete systems. The spread of chloride ions in the concrete causes the steel reinforcement to corrode, which then generates a prevalence of cracks and ultimately, the reinforcing steel and the concrete will split. Consequently, solutions to address this issue are the subject of much research. Studies have shown that the presence of NPs in the concretes can substantially hamper the carbonation (the negative impacts of chlorine ions), thereby protecting the steel structure against oxidation. According to Said et al. [56], the addition of 6% NS to pure cement concrete decreased the depth of chloride penetration by 59.2%–69.6% (10.3, 3.1 and 4.6 mm average penetration depth for 0%, 3% and 6% NS additives to pure cement concrete, respectively, and 8.1, 3.1 and 3.3 mm for 0%, 3% and 6% NS additives to 30% cement substitute fly ash).

Research conducted by Madandoust et al. [97] established that the inclusion of nano-additives reduced the chloride permeability to the extent that it was categorised as low [125]. In comparison to the control sample, the inclusion of 3% of SiO_2, 2% of Fe_2O_3 and 4% of CuO resulted in a decrease in chloride permeability to 60%, 44% and 44%, respectively. Lee et al. [126] examined the penetration of chloride ions in mortar with the inclusion of small-combined additives of CNT-NS. The researchers fixed NS at 1% and altered the CNT levels (0%, 01%, 0.03%, 0.05% and 0.07%). When the 1% NS was added, it decreased the chloride ion penetration from moderate to low (charge passed from 2577 to 1333 coulombs). Moreover, when combined CNT-NS was included in the mortar, the penetration of chloride ions reduced to a very low classification across all specimens. The specimen with 1% NS + 0.03% CNT produced the most significant impact, with the smallest charge passed (457 coulombs), which was approximately 5.64 times lower than the control sample. The major impact that the inclusion of CNT has on the penetration of chloride ions is possibly due to the role played by the C-S-H gel in filling the micropores and bridging the capillary pores in the matrix [127]. Numerous studies have investigated the impacts of 0D nanoparticles (SiO_2, Al_2O_3, ZrO_2, CuO, Fe_2O_3 and $CaCO_3$), both separately and as an amalgamation (Table 2.4). The vast majority of these studies confirmed that chloride resistance penetration is improved with the addition of nanoparticles in small quantities (Figure 2.7). This was largely attributed to the nanoparticles' ability to reduce the permeability and porosity of the concrete/mortar matrix, which is because of their capacity to fill pores and refine microcracks [64,97,128]. Similarly, sulphate attack is a form of chemical attack that negatively impacts the robustness of reinforced concrete and has been explained as chains of the complex chemical reaction between SO_4^{2-} and cement hydration products [129]. The sulphate ions react with these products as gypsum to produce ettringite ($3CaO•Al_2O_3•3CaSO_4•32H_2O$). Ettringite compound formation causes the concrete matrix to expand and crack because of the pressure caused by crystal development [130], swelling [131], raised solids' volume [132] and topochemical reactions [133].

Expansion and cracks cause concrete to lose strength, as does crumbling and weakening of the cementitious mass as time passes [144]. The majority of results so far (Table 2.4) showed that including diverse NPs aids in decreasing sulphate and chloride ion attacks in concrete. This is due to the NPs strong reactivity and wide surface area that improve the $Ca(OH)_2$ reaction, thereby improving the concretes' resistance against different chemical attacks. As previously mentioned, this is largely because of the capacity of NPs to fill gaps and pores in the C-S-H gel. Accordingly, these NPs function as kernels that refine the micropores and cracks of the concrete matrix, thereby enhancing the resistance against sulphate and chloride ion attacks [91,145,146].

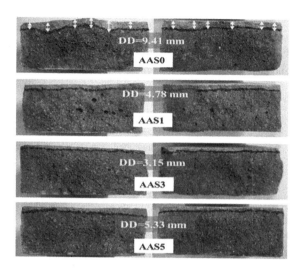

FIGURE 2.7 A cross-sectional images of the sample showing the chloride penetration depths after RCM test (24 hours and 10% of NaCl) [100]. (AAS0, AAS1, AAS3 and AAS5 denotes the alkali-activated slag mortar with 0, 0.01, 0.03 and 0.05 of TiO_2-ghraphine nanosheets, respectively.)

Currently, there is an industrial revolution ongoing in the construction industry wherein construction materials incorporating nanoscience and nanotechnology are creating sustainable economic development. Evidence shows that nanomaterial-based concretes are beneficial for environmental sustainability in the future, as they are energy efficient, produce clean energy, provide strong and robust performances, have improved resistance against arrange of acid attacks, and are eco-friendly in that they are both self-cleaning and self-healing [147–149]. Sustainable development is a key issue in the 21st century, wherein nanomaterial-based concretes can make a positive contribution to producing new eco-friendly building products, thus facilitating a comprehensive understanding of how nano-concretes interact with the environment. Hamers [150] asserts that comprehending the design principles of various nanomaterials from the physicochemical interactions at the molecular scale is extremely important. Furthermore, chemistry knowledge and power must be harnessed to ensure that diverse nanomaterial-based products and technologies for the construction industry are highly environmentally-friendly and have superior economic benefits to the conventional OPC-based products. Essentially, construction nanomaterials that are derived from nanotechnology are potentially advantageous due to their strength, durability and sustainability.

2.5 THE INTERFACIAL TRANSITION ZONE (ITZ)

The interfacial transition zone (ITZ) lies between cement mortars and aggregates and is the weakest area of the concrete [151–153]; thus, it significantly impacts the

TABLE 2.4

Effects of Various Nanomaterials on the Durability Indices of Concretes

NPs	Optimum, %	Finding	Refs.
Al_2O_3	1	Reduction in porosity with addition of nano-Al_2O_3 at all curing conditions The maximum reduction in the porosity was 8.56% for 1% NA replacement cement	[67]
SiO_2	1	When nano-SiO_2, hydrophilic nanomontmorillonite and hydrophobic nanomontmorillonite were admixed into fresh cement mortar at 1% by weight of cement, the value of D_{cl}—was decreased by 61.7%, 66.4% and 76.0%	[134]
SiO_2	3	The reduced depth of carbonation was 73% after 180 days compared to its control samples	[135]
SiO_2	2	Increased sulphuric acid resistance	[70]
TiO_2	1, 3 and 5	The decrease in carbonation depths by addition of 1%, 3% and 5% nano-TiO_2 particles are 77%, 62% and 42% at 180 days compared to the control samples, respectively Reduce the accumulative pore volume	[64]
TiO_2	3	High resistance to deterioration of acid attacked with adding 3% TiO_2 Reduction by 20%–35.5% in the total charge passed at 28 days when increasing the TiO_2 NPs content in the mix from 1% to 5%	[95]
Al_2O_3	3	The reduction in chloride penetration by around 29% and 19%, respectively, at 14 and 28 days, compared to the reference specimen	[136]
SiO_2	2	2% of NS (36.5 nm in size) improved chloride migration coefficients by 70.	[137]
TiO_2	5	Decreased shrinkage Increased carbonation resistance	[64]
TiO_2	0.03	The chloride penetration was reduced by 66.5% compared by its control samples—the results showed lower porosity	[100]
TiO_2	1	The reduction of porosity extent to 16.89% compared to control samples The reduced extent of total specific pore volume reached 11.99% The reduced extent if most probable pore diameter was 31.91 nm The chloride penetration reduced by 31% compared to control samples	[102]
SiO_2	1	The reduction of porosity extent to 6.22% compared to control samples The chloride penetration reduced by 18.04% compared to control samples	
Al_2O_3	1.5	Reduction in chloride penetration by 70%	[104]
ZrO_2	1.5	Reduction in chloride penetration by 80%	
SiO_2	0.25	The maximum reduction in carbonation depth was 47% The maximum reduction in chloride migration depth was 28%	[128]

(Continued)

TABLE 2.4 (*Continued*)
Effects of Various Nanomaterials on the Durability Indices of Concretes

NPs	Optimum, %	Finding	Refs.
SiO_2	2	The reduction in sorptivity coefficients was about 65% compared to reference matrix Enhanced the porosity Enhanced the resistance to attack by sulphuric acid	[70]
SiO_2	3	Decrease the chloride penetration by 66% compared to control samples	[97]
Fe_2O_3	2	Decrease the chloride penetration by 44% compared to control samples	
CuO	4	Decrease the chloride penetration by 44% compared to control samples	
SiO_2	3	Decrease the carbonation depth by 46.47% at 7 days ages Decrease the carbonation depth by 17.42% at 70 days ages	[138]
SiO_2	4	The maximum reduction in chloride penetration depth was 40%	[50]
Nanoclay	4	The maximum reduction rate of porosity was 43% compared to control samples	[139]
Combined of SF with SiO_2	10+2	The maximum reduction of chloride ion penetration was 62%	[79]
SiO_2	3	Increased resistance of sulfphuric acid, magnesium sulphate and seawater (NaCl) solutions	[140]
SiO_2		Decreased carbonation resistance	[141]
SiO_2	10	Improve the resistance to sulphate attack better than control sample	[142]
$CaCO_3$	1	Reduction in chloride diffusion coefficient by about 73% compared to control sample Reducing the water sorptivity by about 17% and 30% compared to cement concrete sample at 28 and 90 days, respectively Reducing the permeable voids by 46% at 28 days compared to cement concrete sample Reducing the chloride ion permeability by 20% and 50% at 28 and 90 days, respectively, compared to cement concrete sample	[143]
SiO_2, Fe_2O_3 and CuO individually	-	Reduce the water absorption by 6%, 4% and 9% for 4 wt% of SiO_2, 3 wt% of Fe_2O_3 and 3 wt% of CuO, respectively The reduction of the chloride permeability was 60%, 44% and 44%, by addition of 3 wt% of SiO_2, 2 wt% of Fe_2O_3 and 4 wt% of CuO, respectively	[97]
TiO_2- graphene nanosheets	0.03	Reduced the coefficient of capillary sorptivity of mortar by 32.0% Reduced the chloride penetration depth by 66.5%	[100]

concretes' properties. A layer of "water film" [154] covers the aggregate during the concrete mixing process, causing the water–cement (w/c) ratios of the aggregate surface and the cement mortar to differ. Large hexagonally shaped flakes of $Ca(OH)_2$ crystals form without difficulty on the surface of the aggregate, growing from the cement mortar surface to the aggregate surface. Some of the properties of the ITZ are that it has greater porosity, less rigidity and higher CH content than the interior cement mortar [155]. ITZs contain numerous pores or small cracks, meaning that seawater can penetrate through this zone into the concrete without much resistance. A reaction occurs between a range of ions (including Ca^{2+}, Na^+, Cl^-, SO_4^{2-}, Mg^{2+} and H^+) in water and loose CH, thereby creating both swelling salts and soluble salts [156]. This causes its structure to be less compact, which is a state linked with wet–dry and freeze–thaw cycles, earthquakes and other conditions specific to the marine environment. Several studies have demonstrated that the presence of nanomaterials can substantially enhance the interface between aggregates and cement mortars [157]. However, unlike ordinary concrete which can be vibrated after pouring, non-dispersible underwater concrete depends solely on its own fluidity to finish the process of compacting after pouring. Achieving a bond strength level comparable to that of ordinary concrete is highly challenging for non-dispersible underwater concrete. Consequently, applying suitable approaches to strengthening the ITZ of non-dispersible underwater concrete is of great importance.

Figures 2.8 and 2.9 depict the three forms of nanomaterials that contribute to enhancing ITZ's bond strength in non-dispersible underwater concrete. The concrete in the early stage (3 days) had a far more evident nanomaterial bond strength than that of the late hydration stage (28 days). As shown in Figure 2.8, nano-SiO_2 most significantly enhanced the concrete. At 3 days, the bonding strength of NS-3 was 134.12% greater than that of PC, whereas NS-1 did not materially impact the

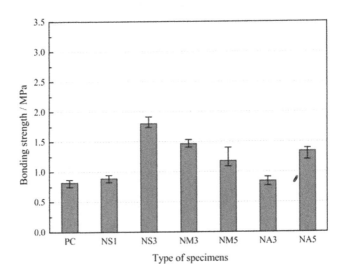

FIGURE 2.8   The bond strength of ITZ at 3 days [153].

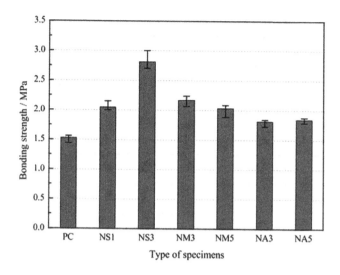

FIGURE 2.9 The bond strength of ITZ at 28 days [153].

bonding strength of the concrete, which is attributable to its minimal nano-SiO_2 content. The addition of nanomaterial clearly improved the bonding strength. The bonding strengths of NM3 and NM5 increased by 87.7% and 43.8%, respectively, over that of PC. Similarly, the addition of nano-Al_2O_3 improved the bonding strengths of NA3 and NA5 by 3.4% and 64.17%, respectively, over that of PC. Figure 2.9 shows that of the three nanomaterials, the bond strength of nano-SiO_2 was superior. At 28 days, the bond strengths of NS1 and NS3 improved by 38.87% and 108.54%, respectively, over that of PC. Moreover, with the addition of nano-metakaolin, the bonding strengths of NM3 and NM5 increased by 43.41% and 31.08%, respectively. When the nano-metakaolin content exceeded 3%, agglomeration occurred, which was detrimental to the strength [158]. There was a minor improvement in the bonding strength of the nano-Al_2O_3 integrated concrete, while there was an increase of approximately 20% for both NA3 and NA5.

The nanomaterials can stimulate CH production in the ITZ during the early stage of hydration. The nanomaterials do not show activity until the CH concentration reaches a particular level. At that point, the crystallised CH of hexagonal slices grows in the ITZ allowing nanomaterials to fill the gaps [159], thereby enhancing the ITZ bond strength at 3 days. The nanomaterials progressively demonstrated pozzolanic activity as the CH concentration continuously increases, and the CH reaction in the ITZ created C-S-H gel or C-A-S-H gel [160,161]. With the progression of the hydration process, the gel wrapped the nanomaterials and gave the reaction formula. During the hydrations' later stage, the network gel produced from the pozzolanic effect steadily filled the ITZ (Figure 2.10), ultimately leading to substantially improved bond strength at 28 days [162].

When nano-SiO_2 was added into the cement-based materials, nano-SiO_2 in ITZ area became "seed" in the early hydration age and promoted suspended the cement

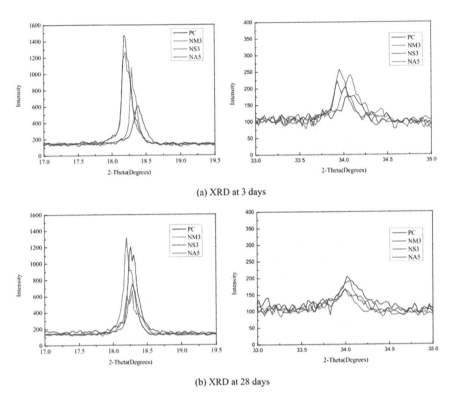

(a) XRD at 3 days

(b) XRD at 28 days

FIGURE 2.10   The XRD value of CH of specimen [153].

particle formation of CH and C-S-H gel. When the CH level became highly concentrated, the nano-SiO$_2$ was very active and reacted with CH, thus producing extra C-S-H gel that filled the ITZ area. In cement with no added nanomaterials, the hydration process was only possible around the cement particles. As nanomaterials were dispersed in the "water film", the ITZ could be directly filled by the hydration products, thus explaining how nanomaterials could improve the strength of the ITZ of NUC [163]. Nano-metakaolin was less active than nano-SiO$_2$ for the pozzolanic effect (Figure 2.11). Furthermore, the vast majority of nano-SiO$_2$ products were C-S-H gel, and following the reaction between the nano-metakaolin and CH, the products were more varied (for instance, C-S-H gel, C$_2$ASH$_8$, C$_4$AH$_{13}$ which was a metastable phenomenon at the early hydration stage due to the supersaturation of the aqueous phase in relation to CH) and hydrogarnet. Nano-metakaolin's multifaceted product creation and low rate of hydration caused the bond strength of ITZ with nano-metakaolin to be lower than that of ITZ with nano-SiO$_2$ [164–166]. Conversely, nano-Al$_2$O$_3$ presented diverse characteristics. Firstly, while nano-Al$_2$O$_3$ was capable of stimulating CH generation, it had very low dissolution. Secondly, even in high-alkalinity pore cements, its pozzolanic effect was a lengthy process; therefore, prior to the 28 days, the ITZ with nano-Al$_2$O$_3$ had low levels of C-S-H gel.

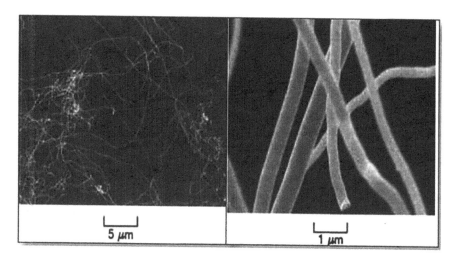

FIGURE 2.11   A clump of VGCNF [167].

According to the push-out test results, nano-SiO$_2$ and nano-metakaolin were highly effective in improving the bond strength of non-dispersible underwater concrete. At 3 days, adding 3% of each enhanced the bond strength of non-dispersible underwater concrete by 134.12% in the case of nano-SiO$_2$ and 87.7% in the case of nano-metakaolin. At 28 days, 3% nano-SiO$_2$ and 3% nano-metakaolin resulted in bond strength of non-dispersible underwater concrete increases of 108.54% and 43.41%, respectively. These results confirm that the addition of either of these two nanomaterials could decrease or prevent a huge binder loss during the process of pouring non-dispersible underwater concrete. In relation to nanomaterials, the ITZ's modulus could be substantially increased while decreasing the width. During the nano indentation test, adding 3% nano-SiO$_2$ and 3% nano-metakaolin caused improvements of 71.03% and 54.39%, respectively, in the ITZ's minimum modulus. All three nanomaterials speeded up the formation of crystals, which created substantial strength in the early stage of hydration. However, when the calcium hydroxide concentration reached a particular level, nano-SiO$_2$ and nano-metakaolin demonstrated clear pozzolanic effects. Additionally, they decreased the calcium hydroxide crystal content in the ITZ and created numerous amorphous gel materials in the later hydration stage. During hydration, the nanomaterials demonstrated diverse functions (e.g., nucleation, filling or pozzolanic effects), which improved the ITZ microstructure, reduced the level of porosity, and densified the ITZ.

2.6   CARBON NANOMATERIALS APPLIED IN CEMENTITIOUS COMPOSITES

The most commonly utilised materials in infrastructure are cementitious composites [167–169]. There are numerous reasons for this, including that the materials are water-resistant, there is no difficulty in terms of the shapes and sizes that can be formed, they are cost-efficient and there is no lack of availability. Cementitious composites are used globally 100% more than the total of all other construction materials, which includes aluminium, steel, wood and plastic [169]. While cement-based composite materials (including concrete and its numerous offshoots) are effective in terms of mechanical performance in compression, these materials typically demonstrate low tensile and flexural strength. Cement composites tend to be quasi-brittle and consequently cracks are prevalent and occur due to tensile stresses [170,171]. One means of addressing this weakness on a macro level is the integration of diverse fibres during the mixing process that positively impacts the mechanical performance in the long term. Various kinds of fibres (including glass, steel and carbon) can be added to the mixture to enhance its resistance to crack development during its lifetime. While they can delay the onset of microcracks, they are incapable of stopping their instigation [172–174]. This is because there is substantial spacing between fibres, meaning that microcracks can freely develop within that space [175,176]. Accordingly, nanofibres and their roles in restricting the onset of cracking at the nanoscale have become a significant topic of interest for researchers. The most widely utilised nanomaterials are nanoscale spherical particles (such as nano-SiO_2, TiO_2, Al_2O_3 and Fe_2O_3), nanotubes and fibres (CNTs and CNFs) and nanoplatelets (nano clays, graphene and graphite oxide) [144].

There are three key benefits to the utilisation of nanomaterials. Firstly, high-strength cementitious composites, and thus high-strength concrete, are produced for a specific purpose. Secondly, the quantity of cement required in concrete to reach comparable strength levels is decreased, as are the costs of construction materials and their impact on the environment. Thirdly, construction is faster as nanomaterials facilitate the production of high-strength concrete with more rapid curing [47]. Numerous research projects have examined nanomaterials, as their particular chemical, electrical, mechanical and thermal properties and effectiveness as reinforcement materials have made them a popular topic [12]. Nanoparticles can have several functions: as heterogeneous nuclei for cement pastes (their high reactivity speeds up the cement hydration process), as nano reinforcement and as nano-filler increasing the density of the microstructure, resulting in decreased porosity [177].

CNFs have an excellent modulus of elasticity and tensile strength in the range of tera pascals (TPa) and giga pascals (GPa), respectively, and they offer distinctive chemical and electronic properties [12]. However, Hogancamp et al. demonstrated that CNFs do not significantly affect cement mortars' stiffness [178]. Figure 2.11 illustrates a clump of vapour-grown CNF (VGCNF). In 1991, Iijima, a Japanese scientist, observed the graphite product via arc evaporation and thereby discovered CNTs, which are allotropes of carbon with a cylindrical nanostructure [179]. There are two primary classifications of CNTs: (i) single-wall CNTs (SWCNTs), which are one graphene sheet rolled into a hollow cylinder, and (ii) multiwalled CNTs (MWCNTs), which are comprised of numerous concentric graphene cylinders arranged together around a hollow core [179,180]. Figure 2.4 presents the structure of the CNTs. CNTs can be synthesised using a variety of techniques including laser ablation, electric arc discharge, solar energy for vaporisation and chemical vapour deposition (CVD) [181]. The electric arc discharge and laser ablation techniques have substantial energy requirements [182]. CVD is typically acknowledged as being a straightforward and efficient CNT production process [183]. Nanotubes develop as a source of gaseous carbon, and this typically entails a hydrocarbon decomposing on the catalyst particles, thereby creating graphitic carbons within the temperature range of 873°C–1273°C [181].

CNTs have a high aspect ratio; hence, crack dissemination requires more energy than low aspect ratio fibre [184]. Moreover, CNTs have exceptional physical traits, including high strength and Young's modulus, which exceed carbon fibres' strength 20-fold and 10-fold [185]. There are two key issues that arise when utilising CNTs. Firstly, the dispersion of fibres in the cement paste, and the bonding between cementitious materials and CNTs. Because of intense van der Waals' influences, CNTs typically produce agglomerates or bundles that have a high probability of defecting sites in the composites [186]. Figure 2.12 depicts the cement-based hybrid material comprising CNTs appended to cement particles [187]. The second issue that arises is the bonding between the cement matrix and CNTs. This is due to few interfacial areas between CNTs and the cement matrix which causes CNTs to be pulled without difficulty from the matrix when under tensile stresses [186].

The intense van der Waals force of nanomaterials means that their distribution in cement composite is extremely challenging. A consequence of this force is that agglomerations form, meaning it becomes highly entangled [188]. Figure 2.13 illustrates a weak interfacial bond between the CNFs and matrix in various regions of the material with numerous cavities. Expanding on past research (discussed earlier) into the importance of nanomaterial length, Yazdanbakhsh et al. examined how dispersion is impacted by length. Cement particles are very large in comparison to nanomaterials' size, which creates problems in relation to size compatibility. This is because nanomaterials are not capable of penetrating the cement grains' area. Consequently, several other areas have a higher nanomaterials' concentration in which nanomaterials' clumping is typical, causing weak dispersion [188]. Furthermore, this weak

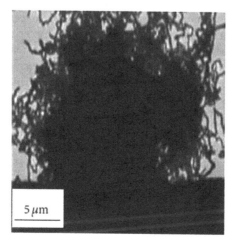

FIGURE 2.12   The TEM image of complete coverage of cement particles by MWCNTs [167].

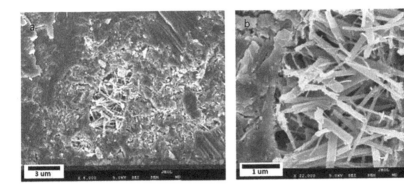

FIGURE 2.13 A clump of CNFs within a pore in the hardened cement paste, and both pictures are taken from the same location in two different scales [167].

dispersion creates unreinforced areas that facilitate the propagation of cracks. The outcome is that weak dispersion can negatively impact the cement composites' mechanical properties and reduce the compressive, flexural and tensile strength. Therefore, dispersion is highly significant to cement composites' mechanical properties.

2.7 NANOMATERIALS-BASED GEOPOLYMER CONCRETE

It is widely acknowledged that Portland cement production is one of the primary greenhouse gas emitters globally. It accounts for between 5% and 8% of all CO_2 emissions to the atmosphere [189–191]. Of the materials utilised by the construction industry, cement is responsible for 36% of emissions and 8% of all anthropogenic CO_2 emissions. At least 70% of greenhouse gas emissions from the production of concrete can be attributed to the production of cement [192]. Hence, all countries have a responsibility to address this issue by discussing regulations for and reduction of CO_2 emissions [193]. Moreover, manufacturing 1 tonne of cement requires approximately 2.8 tonnes of raw materials; therefore, it is a process in which a vast quantity of natural resources are consumed (for instance, limestone and shale required for cement clinkers' production) [194]. In addition, the concrete industry consumes circa 1 trillion litres of mixing water annually [195]. Thus, in order to sustain economic development, it is critical that both renewable and non-renewable materials are applied efficiently and effectively [196]. With the ongoing and severe environmental issues being experienced worldwide, establishing the sustainable development of a replacement for Portland cement (OPC) has become a key priority [197].

Geopolymer technology is an expedient replacement for traditional concrete. It was developed by Davidovits in 1970 in France [198], but its use dates back to Ancient Rome, when castles and monuments were built using geopolymer [199]. Geopolymers are an inorganic aluminosilicate polymer family, synthesised via the alkaline activation of diverse aluminosilicate materials or other industrial or agricultural by-products that are rich in silicon and aluminium, including fly ash (FA), ground blast furnace slag (GBFS), metakaolin (MK), palm oil fuel ash (POFA) and rice husk ash (RHA) [7]. A polymerisation model comparable to that suggested for MKs' alkali activation details the creation of zeolites or zeolite precursors from solutions of alkali aluminosilicate [200]. Polymerisation refers to the chemical reaction that occurs between the alkaline solution and the source binder material. The outcome is a 3D polymeric chain and ring structure Si-O-Al-O bonds [201].

Moreover, the aluminosilicate chains could present as: (i) poly-(sialate), wherein the Si to Al ratio equals 1.0 (-Al-O-Si-chain); (ii) poly (sialate-siloxo), wherein the Si to Al ratio equals 2.0 (-Al-O-Si-Si-chain); and (iii) poly (sialate-disiloxo), wherein the Si to Al ratio equals 3.0 (-Al-O-Si-Si-Si-chain) [202].

Fundamentally, nanotechnology is the capacity to observe and reorganise matter at the atomic and molecular levels within the range of 1–100 nm, in addition to contributing the unique traits and phenomenon at that size that are comparable to those associated with single atoms and molecules or bulk behaviour [147]. The topic of nanotechnology has become increasingly popular amongst researchers and over the last 20 years, new scientific and practical applications have progressively come to the fore. Lately, there have been significant endeavours to integrate NPs in construction materials in order to boost their properties and produce concrete that performs more effectively [203]. Researchers have given focus to nanocomposite construction materials due to the distinctive physical and chemical characteristics of NPs, which are attributable to their ultrafine-sized particles [204]. NPs such as NS [114], nano-Fe_2O_3 and nano-Fe_3O_4 [205], nano-Al_2O_3 [206], nano-titanium dioxide [207], CNTs [208], nano-$CaCO_3$ [157], nanoclay [209] and multi-wall CNTs [210] have been frequently utilised to improve performances in conventional cement-based concrete composites. Conversely, in order to improve the geopolymer mixtures' durability, physical structure and mechanical traits [211], NPs have been integrated with geopolymer matrices. The larger surface area-to-volume ratio of NPs mean that they are extremely reactive and significantly impact reaction rates [212]. Accordingly, the geopolymer concretes' microstructure is changed at the atomic level by NPs, which generates substantial enhancements in the properties and structural behaviour in both the fresh and hardened states, without the inclusion of heat [93]. However, some research utilised NPs in geopolymer pastes as a functional mixture for non-structural applications including energy storage, self-cleaning, and antibacterial applications.

The existing research describes attempts to enhance different geopolymer composites' properties with the utilisation of diverse nanomaterials including nanoclay platelets (NCP), NS, colloidal NS (CNS), nanoclay (NC), nano-alumina (NA), CNTs, multi-wall CNTs (MWCNTs), nano-titanium (NT), waste glass nano-powder (WGNP), nano-metakaolin (NM), nano-calcium carbonate (NCC), nano-zinc oxide (NZn), NS slurry (NSS) and graphene nanoplatelets (GNP) [213]. However, as evidenced by the data and analysis presented in Figure 2.14, NS is the most commonly utilised nanomaterial in geopolymer composites. Furthermore, this was also the case for cement and traditional concrete composites [214], which is because of NS's effective pore-filling and pozzolanic reactivity. The main component of NS is silicon dioxide (SiO_2), which is available in both crystalline and amorphous states. Typically, NS in its amorphous state was employed to form a variety of concretes [215]. NS is comprised of spherical particles or microspheres with a main diameter of 150 nm and a high specific surface area of 150–250 m²/g that is generated via vaporising

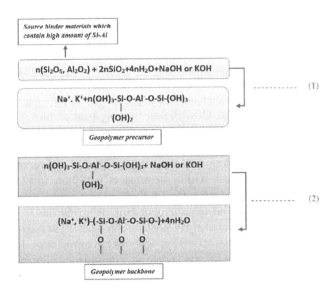

FIGURE 2.14 Chemical reactions during the geopolymerisation process [217].

silica in an electric arc furnace at temperatures ranging from 1500°C to 2000°C by decreasing quartz. In general, NS particle size varied from 5 to 658 nm for diverse NS-based products [216]. There are five key changes and improvements that are the outcome of adding the aforementioned NP types, as follows: (i) the source binder materials' polymerisation and hydration, (ii) greater mechanical properties, (iii) the composites' microstructure is densified, (iv) lower water permeability and (v) more resistant to extreme environmental conditions.

Researchers have examined the fresh, mechanical and microstructural properties of FA/GBFS-based geopolymer concrete integrated with NS at diverse volumes. Evidence shows that when compared to the original mixture with no NS, all mixtures integrated with NS had higher compressive strength (CS). It was found that 1.5% of NS additive produced the highest CS, increasing it by 11% above the control mix after 28 days of ambient curing temperature. The strength improvement is due to the silica nanoparticles filling the nanopores inside the geopolymer concrete, thereby increasing the compactness and density of the matrix. Moreover, NS's chemical composition, which has high levels of silica, speeds up the geopolymer reactions and strengthens their geopolymer binder, which ultimately improves the specimens' strength. Furthermore, 1.5% NS is considered the optimal content level to enhance CS, as exceeding this level causes a minimal decrease in CS as the numerous unreacted NS particles in the matrix coupled with the excess NS causes agglomerations to form between the NS particles that could otherwise have prevented the dissolution of the silica [218]. This reduced geopolymer concrete CS subsequently causes voids to appear [218]. Many studies have corroborated that the CS of geopolymer concrete is enhanced by the integration of NS [219]. Moreover, Nuaklong et al. [220] determined that integrating up to 2% NS enhances geopolymer concretes' CS but decreases it once it exceeds this level. In contrast, Angelin Lincy and Velkennedy [221] found that the cut-off level for adding NS and seeing an improvement in the CS of geopolymer concrete was 0.5%. However, according to Ibrahim et al. [222] and Janaki et al. [223], there was actually a 5% threshold for improvement in the CS of geopolymer concrete with the addition of NP, after which it declined. CS increases of 1.5%, 13.6% and 1.3% were observed at 2%, 5% and 10% of CNT, respectively, at 28 days [223], whereas the CS increase was recorded as 0%, 8.2%, 23.3% and 19.8% at 1%, 2.5%, 5% and 7.5%, respectively, of NS, also at 28 days [222]. Contrastingly, Kotop et al. [224] found maximum CS to be at the level of 2.5% NC added to geopolymer concrete mixtures. In line with this, improvements of 90% and 70% in the CS over the virgin mix were recorded at 28 and 60 days, respectively. It was also found that CS decreased after 2.5% of NC. This decrease was attributable to the agglomeration of the NC and weak dispersion inside the concrete mixture [225]. In addition, when compared to the control mixes, when 0.02% of CNTs were added to the geopolymer concrete mixtures, improvements of approximately 81% and 57% in the CS were determined at 28- and 60-day curing, respectively [224]. The researchers maintained that this increase in strength was due to the Cants essentially functioning as bridges for decreasing the spread of both micro and macro cracks [226]. The alkaline liquid has an impact on the dispersion of Cants as they were surfactant by sodium hydroxide, thereby enabling them to generate effectively dispersed Cants and de-bundled within the geopolymer concrete matrix [227]. Similarly, when compared to the control geopolymer concrete mixture that did not include any NPs, Kotop et al. [224] recorded top CS enhancements of 99% and 70% at 28- and 60-day curing, respectively, for the geopolymer concrete mixtures integrated with a mix of 2.5% NS and 0.01% Cants.

2.8 EFFECTS OF NANOMATERIALS ON ALKALI-ACTIVATED BINDERS

A number of researchers have also found that adding nanomaterials to the geopolymeric binder improved the mechanical properties. The theory behind this improvement has three core aspects: (i) nano filling impact, (ii) hydration through nucleation sites and (iii) bridging of cracks. The following section offers a comprehensive overview of the latest findings about how the mechanical properties of alkali-activated binders are impacted by the addition of the following nanomaterials: nano-SiO_2, nano-Al_2O_3, nano-TiO_2 and CNTs.

NS is a highly reactive and fine pozzolanic material. There are various means of producing it, including the sol-gel process [228], thermal decomposition technology [229] and vapour phase [230]. Due to its strong pozzolanic reactivity and its pore-filling effect, the use of NS is prevalent in cementitious mixtures, as it facilitates enhanced concrete hardening [231]. Nano-SiO_2 has a very fine spherical texture, with an average diameter of 4 nm and a comparatively high surface area-to-volume ratio. These characteristics are likely the cause of its high reactivity in the geopolymerisation reactions [72]. Recent research has examined the utilisation of NS in alkali-activated binders and has determined that it is pivotal in speeding up the hydration and polymerisation reactions and also improves the mechanical properties and the characteristics of the microstructure [232]. In summary, NS has three key impacts on alkali-activated binders: (i) faster hydration, (ii) refined microstructures and (iii) improved mechanical properties [70].

In geopolymer, the most commonly utilised nanomaterial is nano-SiO$_2$ [233]. Research has shown that integrating 1% of NS with metakaolin geopolymers resulted in a substantial increase of 52% in the CS at 60 days; however, increasing the additive to 3% caused a decrease in CS [234]. This decline in CS was due to nanoparticles that were unreacted and produced nano-interaction energy and weak dispersion. Another study found that the optimal level of NS to be integrated with coal fly ash/slag bend-based geopolymer is approximately 2% (Figure 2.15a). However, it was also found that the addition of NS caused a minor delay in the hydration and additional calcium-alumina-silicate-hydrates gel was noted, with decreased porosity of the geopolymer [76]. Moreover, it was determined that NS improves the extent of geopolymerisation, ultimately generating a denser geopolymer paste [218]. Figure 2.15b shows the impact of NS on the CS of coal ash/slag geopolymer. Another recent research project examined the integration of diverse levels of NS (1%–8%) with geopolymers and found that CS increased up to 3% NS addition, but exceeding this level caused nano-agglomeration and was detrimental to the geopolymeric gels' mechanical properties [235]. Others maintained that employing the wet-mix method of adding NS was a more successful means of enhancing the microstructures and mechanical traits of geopolymers [236].

The widely accepted theory behind this improvement refers to enabling geopolymerisation and stimulating denser geopolymer gel development. It was established that NS pozzolanic effect was not the sole reason for alkaline-activated binders' strength improvement. Nanoparticles' filler effect improves the cohesion of the alkali-activated microstructure, thereby enhancing the aggregate-paste bonding [232]. Furthermore, the NS provided more amorphous silicate to instigate geopolymerisation, which is the core source of strength in polymers. All of this is surplus to the theory of the nucleation impact of nanoparticles in the creation of aluminosilicate gels [93].

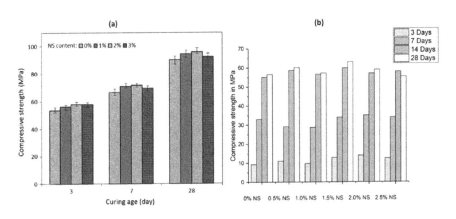

FIGURE 2.15 Effect of nano-SiO$_2$ on the compressive strength of (a) slag/fly ash geopolymer [236] and (b) slag-based geopolymers [218].

It has been suggested that nano-alumina would be a suitable additive for cement and concrete because of its thermal stability, high surface area and unique mechanical characteristics [237]. Nano-Al_2O_3 is nanosized material utilised in cement-based materials. It is advantageous due to its refined microstructure, that is, its pore composition and ITZ [77]. Nano-Al_2O_3's impact on mechanical characteristics is chiefly due to its physical (filler) effect, resulting in an accelerated early-stage hydration [238]. Some researchers have demonstrated that nano-Al_2O_3 has no or only minimal positive impact on the strength of concrete [239], which is likely due to weak dispersion of nanoparticles in cement-based materials.

Recently, there has been significant focus on the impact of nano-alumina on geopolymers [240]. Reports have stated that utilising nano-Al_2O_3 as an additive decreases the setting time required for high-calcium FA geopolymeric binders, while only an insignificant improvement in the strength level was noted [71]. In another study, Phoo-ngernkham et al. [41] found that adding nano-Al_2O_3 into high-calcium FA at 1%, 2% and 3%, improved the compressive and flexural strength of the room temperature–cured specimens. The researchers attribute the improvement to the generation of extra geopolymerisation and hydration products [41].

As illustrated in Figure 2.16, recent research reported that employing nano-Al_2O_3 at 2% as an additive to slag-based geopolymer enhanced the compressive strength at all stages up to 120 days [241]. A correlation was identified between the contribution made by the nano-Al_2O_3 in the aluminosilicate formation and the increased compressive strength. In contrast, utilising a nano-Al_2O_3 additive that exceeds 3% reduces the compressive strength at 7-, 28- and 90-day curing. This declining strength was linked with the inadequate dispersion of nano particles and the formation of weak zones within the geopolymer matrix.

The properties of NT dioxide include that it is fine and has spherical or ellipsoidal powder-based materials and diameters under 100 nm [242]. Nano-TiO_2 is becoming increasingly popular in relation to cementitious materials, as it can offer distinctive traits for architectural constructions. Titanium dioxide (TiO_2) is capable of decomposing a vast array of organic and inorganic air pollutants under ultraviolet light and humidity; therefore, it generates cleaner air and enhanced well-being for people and

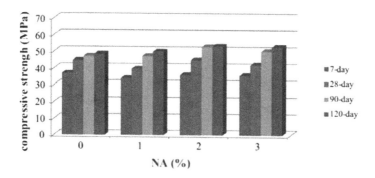

FIGURE 2.16 Compressive strength of alkali-activated slag without and with the addition of nano-Al_2O_3 [241].

other living creatures. Another advantage is that the white colour of TiO$_2$ cement/concrete usually has broad appeal architecturally [243]. Evidence shows that the integration of titanium in construction materials generates cleaner air and "self-cleaning" properties that help to eradicate detrimental bacteria [244]. Moreover, using NT as an additive in concrete mixtures speeds up the cement hydration process and increases the concretes' durability as water permeability is decreased [245]. Photocatalysis refers to the process of photodecomposition via light absorption, and titanium dioxide is recognised as one of the best photocatalysts. The "self-cleaning" nature of TiO$_2$ means that when it is mixed with cement, this advantage can be harnessed on a wide scale; for instance, by enhancing the air quality of urban areas.

There are many references in the existing literature to the mechanism connected with the self-cleaning properties of photocatalytic materials including nano-TiO$_2$ [246]. Ultraviolet (UV) radiation enables organic matter to decompose quickly, thereby enabling the concrete to self-clean. This means that the surfaces of constructions and buildings are cleaner, and there is less surrounding air pollution. In sunlight, nano-TiO$_2$ functions as a photocatalyst to decompose the organic compound. Water will penetrate and remove any dust lying on the surface. Past research has provided evidence of the capacity of nano-TiO$_2$ to decrease pollutants including nitrogen oxides (NOx), aldehydes and volatile organic compounds (VOCs) [101,102]. A study conducted recently by Sanalkumar and Yang [247] examined the ability of metakaolin-based geopolymer integrating nano-TiO$_2$ to self-clean. Recent work published by Sanalkumar and Yang [247] investigated the self-cleaning performance of metakaolin-based geopolymer incorporating nano-TiO$_2$. They found that in addition to the enhancement of the mechanical properties, the total solar reflectance (TSR), photoinduced hydrophilicity and contaminate decomposition have substantially improved with the use of TiO$_2$. Utilising nano-TiO$_2$ as an additive improves the mechanical characteristics of alkali-activated binders by enhancing the microstructure of the geopolymer matrix [247].

Research has shown that substituting FA for 1% and 5% nano-TiO$_2$ improved the coal ash-based geopolymers' compressive strength [248]. Figure 2.17 presents

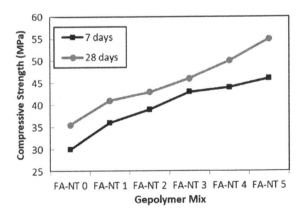

FIGURE 2.17 Effect of nano-TiO$_2$ on the compressive strength of coal fly ash geopolymer [248].

a comparison of the compressive strength for coal ash geopolymer specimens at 7 and 28 days, with and without nano-TiO_2 additive. It was found that increasing the nano-TiO_2 content enhances the CS, with a maximum improvement exceeding 50%. A recent study carried out by Maiti et al. [249] investigated the impact of diverse sizes of nano-TiO_2 (T30, T50, T100) on FA geopolymer synthesised at ambient temperature (Figure 2.18). The researchers discovered that at 28 days of curing, 30 nm nano-TiO_2 (5%) enabled geopolymer with maximum CS surpassing 50 MPa and split tensile strength of approximately 7 MPa (Figure 2.19). Duan et al. [64] noted that strength improves more quickly during the early stages (up to 28 days) than the later stages (up to 90 days). The researchers attributed this improvement in compressive and split tensile strength to the fact that the geopolymerisation process was enabled and a denser microstructure was formed.

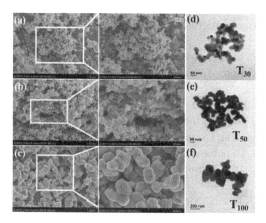

FIGURE 2.18 Effect of nano-TiO_2 size on geopolymer microstructural features [249].

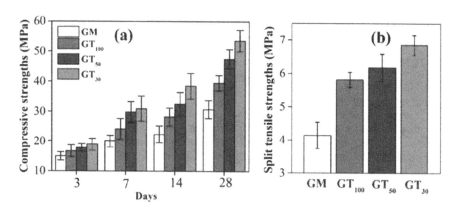

FIGURE 2.19 Mechanical properties of evaluated geopolymers (a) compressive strength (b) splitting tensile strength [249].

2.9 SUMMARY

The mechanisms of effects of nanomaterials on the engineering properties of cement concrete, geopolymer and alkali-activated materials are described in detail in the review. Based on the study review on nanomaterials concrete, the following conclusions were drawn:

i. Incorporated nanoparticles into the mix resulted in a reduction in the time for setting and workability. This reduction was attributed to the high reactivity on nanoparticles. The materials play a noticeable role in improving the strength performance of concrete especially in early ages as a result of the pozzolanic reaction of NPs with $Ca(OH)_2$, which generate more calcium-silicate hydrate C-S-H gel, and filling the pores in the concrete system.

ii. Nanoparticle additives work to enhance the durability by increasing its resistance concrete to carbonisation and chemical attacks due to their role in improving microstructure, refining the microcracks and reducing porosity.

iii. Adding CNTs to the cement composite resulted in better improvements in the mechanical properties of cement composites in comparison with other nanomaterials.

iv. The addition of NS to free cement concrete (geopolymer and alkali-activated) was found to promote the geopolymerisation reactions, shortening the set time and enhancing the mechanical and durability of the resulting geopolymeric binder. To achieve the desired mechanical properties and durability characteristics, the optimum dosage of NS was found to be in the range of 2–3 wt%.

v. The use of CNTs was found to also enhance the geopolymerisation by offering additional nucleation sites. The optimum dosage of CNTs was found to be about 0.1%.

vi. Increasing the percentage of nanomaterial dosage by more than the optimum percentage reflected negatively on the properties of strength and durability due to the difficulty of uniform dispersion and the formation of weak spots within the matrix.

REFERENCES

1. Monteiro, P.J.M., S.A. Miller, and A. Horvath. Towards sustainable concrete. *Nature Materials*, 2017, **16**(7): pp. 698–699.
2. Samadi, M., et al. Waste ceramic as low cost and eco-friendly materials in the production of sustainable mortars. *Journal of Cleaner Production*, 2020: p. 121825.
3. Onaizi, A.M., et al. Effect of nanomaterials inclusion on sustainability of cement-based concretes: A comprehensive review. *Construction and Building Materials*, 2021, **306**: p. 124850.
4. Sahoo, S., P.K. Parhi, and B. Chandra Panda. Durability properties of concrete with silica fume and rice husk ash. *Cleaner Engineering and Technology*, 2021, **2**: 100067.
5. IEA. *Cement technology roadmap plots path to cutting CO₂ emissions 24% by 2050*, 6 April 2018. Available from: https://www.iea.org/news/cement-technology-roadmap-plots-path-to-cutting-CO₂-emissions-24-by-2050.

6. Cao, Z., et al. Elaborating the history of our cementing societies: An in-use stock perspective. *Environmental Science & Technology*, 2017, **51**(19): pp. 11468–11475.
7. Fahim Huseien, G., et al. Geopolymer mortars as sustainable repair material: A comprehensive review. *Renewable and Sustainable Energy Reviews*, 2017, **80**: pp. 54–74.
8. Costa, F.N. and D.V. Ribeiro. Reduction in CO_2 emissions during production of cement, with partial replacement of traditional raw materials by civil construction waste (CCW). *Journal of Cleaner Production*, 2020, **276**.
9. Belbute, J.M. and A.M. Pereira. Reference forecasts for CO_2 emissions from fossil-fuel combustion and cement production in Portugal. *Energy Policy*, 2020, **144**: p. 111642.
10. Pacyna, E.G., et al. Global emission of mercury to the atmosphere from anthropogenic sources in 2005 and projections to 2020. *Atmospheric Environment*, 2010, **44**(20): pp. 2487–2499.
11. Streets, D., et al. Anthropogenic mercury emissions in China. *Atmospheric Environment*, 2005, **39**(40): pp. 7789–7806.
12. Sanchez, F. and K. Sobolev. Nanotechnology in concrete – A review. *Construction and Building Materials*, 2010, **24**(11): pp. 2060–2071.
13. Abdoli, H., et al. Effect of high energy ball milling on compressibility of nanostructured composite powder. *Powder Metallurgy*, 2011, **54**(1): pp. 24–29.
14. Jankowska, E. and W. Zatorski. Emission of nanosize particles in the process of nanoclay blending. In: *2009 Third International Conference on Quantum, Nano and Micro Technologies*, 2009. United States: IEEE: pp. 147–151.
15. Bhatia, S.. Nanoparticles types, classification, characterization, fabrication methods and drug delivery applications. In: *Natural Polymer Drug Delivery Systems*, 2016. Springer, Switzerland: pp. 33–93.
16. Shah, K.W. and G.F. Huseien. Inorganic nanomaterials for fighting surface and airborne pathogens and viruses. *Nano Express*, 2020, **1**(3): 1–16.
17. Shah, K.W., G.F. Huseien, and H.W. Kua. A state-of-the-art review on core–shell pigments nanostructure preparation and test methods. In: *Micro*, 2021. Multidisciplinary Digital Publishing Institute, 1(1): pp. 55–85.
18. Sadeghi-Nik, A., et al. Modification of microstructure and mechanical properties of cement by nanoparticles through a sustainable development approach. *Construction and Building Materials*, 2017, **155**: pp. 880–891.
19. Pacheco-Torgal, F. and S. Jalali. Nanotechnology: Advantages and drawbacks in the field of construction and building materials. *Construction and Building Materials*, Netherlands, 2011, **25**(2): pp. 582–590.
20. Gajanan, K. and S.N. Tijare. Applications of nanomaterials. *Materials Today: Proceedings*, 2018, **5**(1, Part 1): pp. 1093–1096.
21. Fulekar, M. *Nanotechnology: Importance and Applications*, 2010. IK International Pvt Ltd, New Delhi, India, pp. 1–26.
22. Jones, W., et al. Nanomaterials in construction – What is being used, and where? *Proceedings of the Institution of Civil Engineers – Construction Materials*, 2019, **172**(2): pp. 49–62.
23. Lin, C.-C. and W.-Y. Chen. Effect of paint composition, nano-metal types and substrate on the improvement of biological resistance on paint finished building material. *Building and Environment*, 2017, **117**: pp. 49–59.
24. Shah, K.W. and Y. Lu. Morphology, large scale synthesis and building applications of copper nanomaterials. *Construction and Building Materials*, 2018, **180**: pp. 544–578.
25. Zhang, X., et al. Design of glass fiber reinforced plastics modified with CNT and pre-stretching fabric for potential sports instruments. *Materials & Design*, 2016, **92**: pp. 621–631.
26. Norhasri, M.M., M. Hamidah, and A.M. Fadzil. Applications of using nano material in concrete: A review. *Construction and Building Materials*, 2017, **133**: pp. 91–97.

27. Papadaki, D., G. Kiriakidis, and T. Tsoutsos. Applications of nanotechnology in construction industry. In: *Fundamentals of Nanoparticles*, Elsevier, Netherlands, 2018: pp. 343–370.

28. Kewalramani, M.A. and Z.I. Syed. Application of nanomaterials to enhance microstructure and mechanical properties of concrete. *International Journal of Integrated Engineering*, 2018, **10**(2): 1–12.

29. Rajak, M.A.A., Z.A. Majid, and M. Ismail. Morphological characteristics of hardened cement pastes incorporating nano-palm oil fuel ash. *Procedia Manufacturing*, 2015, **2**: pp. 512–518.

30. Li, G.. Properties of high-volume fly ash concrete incorporating nano-SiO_2. *Cement and Concrete Research*, 2004, **34**(6): pp. 1043–1049.

31. Ghafari, E., et al. The effect of nanosilica addition on flowability, strength and transport properties of ultra high performance concrete. *Materials & Design*, 2014, **59**: pp. 1–9.

32. Seifan, M., S. Mendoza, and A. Berenjian. Mechanical properties and durability performance of fly ash based mortar containing nano- and micro-silica additives. *Construction and Building Materials*, 2020, **252**: p. 119121.

33. Khaloo, A., M.H. Mobini, and P. Hosseini. Influence of different types of nano-SiO_2 particles on properties of high-performance concrete. *Construction and Building Materials*, 2016, **113**: pp. 188–201.

34. Mostafa, S.A., et al. Influence of nanoparticles from waste materials on mechanical properties, durability and microstructure of UHPC. *Materials*, 2020, **13**(20): p. 4530.

35. Adak, D., M. Sarkar, and S. Mandal. Effect of nano-silica on strength and durability of fly ash based geopolymer mortar. *Construction and Building Materials*, 2014, **70**: pp. 453–459.

36. Li, Z., et al. Investigations on the preparation and mechanical properties of the nano-alumina reinforced cement composite. *Materials Letters*, 2006, **60**(3): pp. 356–359.

37. Nazari, A., et al. Influence of Al_2O_3 nanoparticles on the compressive strength and workability of blended concrete. *Journal of American Science*, 2010, **6**(5): pp. 6–9.

38. Nazari, A., et al. Mechanical properties of cement mortar with Al_2O_3 nanoparticles. *Journal of American Science*, 2010, **6**(4): pp. 94–97.

39. Hase, B. and V. Rathi. Properties of high strength concrete incorporating colloidal nano-Al_2O_3. *International Journal of Innovative Research in Science, Engineering and 'Technology*, 2015, **4**(3): pp. 959–963.

40. Behfarnia, K. and N. Salemi. The effects of nano-silica and nano-alumina on frost resistance of normal concrete. *Construction and Building Materials*, 2013, **48**: pp. 580–584.

41. Phoo-ngernkham, T., et al. The effect of adding nano-SiO_2 and nano-Al_2O_3 on properties of high calcium fly ash geopolymer cured at ambient temperature. *Materials & Design*, 2014, **55**: pp. 58–65.

42. Massa, M.A., et al. Synthesis of new antibacterial composite coating for titanium based on highly ordered nanoporous silica and silver nanoparticles. *Materials Science and Engineering: C*, 2014, **45**: pp. 146–153.

43. Li, H., M.-H. Zhang, and J.-P. Ou. Abrasion resistance of concrete containing nano-particles for pavement. *Wear*, 2006, **260**(11–12): pp. 1262–1266.

44. Li, H., M.-H. Zhang, and J.-P. Ou. Flexural fatigue performance of concrete containing nano-particles for pavement. *International Journal of Fatigue*, 2007, **29**(7): pp. 1292–1301.

45. Sorathiya, J., S. Shah, and S. Kacha. Effect on addition of nano "titanium dioxide" (TiO_2) on compressive strength of cementitious concrete. *Kalpa Publications in Civil Engineering*, 2018, **1**: pp. 219–211.

46. Jayapalan, A., B. Lee, and K. Kurtis. Effect of nano-sized titanium dioxide on early age hydration of Portland cement. In: *Nanotechnology in Construction 3*, 2009. Netherlands: Springer: pp. 267–273.

47. Morsy, M., S. Alsayed, and M. Aqel. Hybrid effect of carbon nanotube and nano-clay on physico-mechanical properties of cement mortar. *Construction and Building Materials*, 2011, **25**(1): pp. 145–149.
48. Stynoski, P., P. Mondal, and C. Marsh. Effects of silica additives on fracture properties of carbon nanotube and carbon fiber reinforced Portland cement mortar. *Cement and Concrete Composites*, 2015, **55**: pp. 232–240.
49. Mohamed, A.M.. Influence of nano materials on flexural behavior and compressive strength of concrete. *HBRC Journal*, 2016, **12**(2): pp. 212–225.
50. Beigi, M.H., et al. An experimental survey on combined effects of fibers and nanosilica on the mechanical, rheological, and durability properties of self-compacting concrete. *Materials & Design*, 2013, **50**: pp. 1019–1029.
51. Liu, Z.G., et al. Piezoresistive properties of cement mortar with carbon nanotube. In: *Advanced Materials Research*, 2011. Trans Tech Publ, 286: pp. 310–313.
52. Konsta-Gdoutos, M.S., Z.S. Metaxa, and S.P. Shah. Highly dispersed carbon nanotube reinforced cement based materials. *Cement and Concrete Research*, 2010, **40**(7): pp. 1052–1059.
53. Folli, A., et al. TiO_2 photocatalysis in cementitious systems: Insights into self-cleaning and depollution chemistry. *Cement and Concrete Research*, 2012, **42**(3): pp. 539–548.
54. Mudimela, P.R., et al. Synthesis of carbon nanotubes and nanofibers on silica and cement matrix materials. *Journal of Nanomaterials*, 2009, **2009**. p. 1–5.
55. Zhang, M.-H. and J. Islam. Use of nano-silica to reduce setting time and increase early strength of concretes with high volumes of fly ash or slag. *Construction and Building Materials*, 2012, **29**: pp. 573–580.
56. Said, A.M., et al. Properties of concrete incorporating nano-silica. *Construction and Building Materials*, 2012, **36**: pp. 838–844.
57. Land, G. and D. Stephan. The influence of nano-silica on the hydration of ordinary Portland cement. *Journal of Materials Science*, 2012, **47**(2): pp. 1011–1017.
58. Land, G. and D. Stephan. Controlling cement hydration with nanoparticles. *Cement and Concrete Composites*, 2015, **57**: pp. 64–67.
59. Mukharjee, B.B. and S.V. Barai. Assessment of the influence of nano-silica on the behavior of mortar using factorial design of experiments. *Construction and Building Materials*, 2014, **68**: pp. 416–425.
60. Palla, R., et al. High strength sustainable concrete using silica nanoparticles. *Construction and Building Materials*, 2017, **138**: pp. 285–295.
61. Wang, L., H. Zhang, and Y. Gao. Effect of TiO_2 nanoparticles on physical and mechanical properties of cement at low temperatures. *Advances in Materials Science and Engineering*, 2018, **2018**: pp. 1–12.
62. Liu, J., et al. Effects of zinc oxide nanoparticles on early-age hydration and the mechanical properties of cement paste. *Construction and Building Materials*, 2019, **217**: pp. 352–362.
63. Gopalakrishnan, R. and S. Nithiyanantham. Effect of ZnO nanoparticles on cement mortar for enhancing the physico-chemical, mechanical and related properties. *Advanced Science, Engineering and Medicine*, 2020, **12**(3): pp. 348–355.
64. Duan, P., et al. Effects of adding nano-TiO_2 on compressive strength, drying shrinkage, carbonation and microstructure of fluidized bed fly ash based geopolymer paste. *Construction and Building Materials*, 2016, **106**: pp. 115–125.
65. Kontoleontos, F., et al. Influence of colloidal nanosilica on ultrafine cement hydration: Physicochemical and microstructural characterization. *Construction and Building Materials*, 2012, **35**: pp. 347–360.
66. Mirgozar Langaroudi, M.A. and Y. Mohammadi. Effect of nano-clay on workability, mechanical, and durability properties of self-consolidating concrete containing mineral admixtures. *Construction and Building Materials*, 2018, **191**: pp. 619–634.

67. Nazari, A. and S. Riahi. Improvement compressive strength of concrete in different curing media by Al_2O_3 nanoparticles. *Materials Science and Engineering: A*, 2011, **528**(3): pp. 1183–1191.

68. Nazari, A., et al. Benefits of Fe_2O_3 nanoparticles in concrete mixing matrix. *Journal of American Science*, 2010, **6**(4): pp. 102–106.

69. Nazari, A., et al. The effects of incorporation Fe_2O_3 nanoparticles on tensile and flexural strength of concrete. *Journal of American Science*, 2010, **6**(4): pp. 90–93.

70. Deb, P.S., P.K. Sarker, and S. Barbhuiya. Sorptivity and acid resistance of ambient-cured geopolymer mortars containing nano-silica. *Cement and Concrete Composites*, 2016, **72**: pp. 235–245.

71. Chindaprasirt, P., et al. Effect of SiO_2 and Al_2O_3 on the setting and hardening of high calcium fly ash-based geopolymer systems. *Journal of Materials Science*, 2012, **47**(12): pp. 4876–4883.

72. Aggarwal, P., R.P. Singh, and Y. Aggarwal. Use of nano-silica in cement based materials—A review. *Cogent Engineering*, 2015, **2**(1): p. 1078018.

73. Reches, Y.. Nanoparticles as concrete additives: Review and perspectives. *Construction and Building Materials*, 2018, **175**: pp. 483–495.

74. Li, X., et al. Effects of graphene oxide agglomerates on workability, hydration, microstructure and compressive strength of cement paste. *Construction and Building Materials*, 2017, **145**: pp. 402–410.

75. Gowda, R., et al. Effect of nano-alumina on workability, compressive strength and residual strength at elevated temperature of cement mortar. *Materials Today: Proceedings*, 2017, **4**(11): pp. 12152–12156.

76. Gao, X., Q.L. Yu, and H.J.H. Brouwers. Characterization of alkali activated slag–fly ash blends containing nano-silica. *Construction and Building Materials*, 2015, **98**: pp. 397–406.

77. Nazari, A. and S. Riahi. RETRACTED: Al_2O_3 nanoparticles in concrete and different curing media. *Energy and Buildings*, 2011, **43**(6): pp. 1480–1488.

78. Hosseini, P., A. Booshehrian, and A. Madari. Developing concrete recycling strategies by utilization of nano-SiO_2 particles. *Waste and Biomass Valorization*, 2011, **2**(3): pp. 347–355.

79. Jalal, M., et al. Comparative study on effects of Class F fly ash, nano silica and silica fume on properties of high performance self compacting concrete. *Construction and Building Materials*, 2015, **94**: pp. 90–104.

80. Collepardi, M., et al. Influence of amorphous colloidal silica on the properties of self-compacting concretes. In: *Proceedings of the International Conference "Challenges in Concrete Construction-Innovations and Developments in Concrete Materials and Construction"*, Dundee, Scotland, UK, 2002: p. 1–12.

81. Joshaghani, A., et al. Effects of nano-TiO_2, nano-Al_2O_3, and nano-Fe_2O_3 on rheology, mechanical and durability properties of self-consolidating concrete (SCC): An experimental study. *Construction and Building Materials*, 2020, **245**: p. 118444.

82. Sumesh, M., et al. Incorporation of nano-materials in cement composite and geopolymer based paste and mortar – A review. *Construction and Building Materials*, 2017, **148**: pp. 62–84.

83. Vikulin, V.V., M.K. Alekseev, and I.L. Shkarupa. Study of the effect of some commercially available nanopowders on the strength of concrete based on alumina cement. *Refractories and Industrial Ceramics*, United States, 2011, **52**(4): pp. 288–290.

84. Rashad, A.M.. A synopsis about the effect of nano-Al_2O_3, nano-Fe_2O_3, nano-Fe_3O_4 and nano-clay on some properties of cementitious materials – A short guide for Civil Engineer. *Materials & Design* (1980–2015), 2013, **52**: pp. 143–157.

85. Rashad, A.M.. Effects of ZnO_2, ZrO_2, Cu_2O_3, CuO, $CaCO_3$, SF, FA, cement and geothermal silica waste nanoparticles on properties of cementitious materials – A short guide for Civil Engineer. *Construction and Building Materials*, 2013, **48**: pp. 1120–1133.

86. Zhang, R., et al. Influences of nano-TiO$_2$ on the properties of cement-based materials: Hydration and drying shrinkage. *Construction and Building Materials*, 2015, **81**: pp. 35–41.

87. Wang, L., et al. Effect of nano-SiO$_2$ on the hydration and microstructure of Portland cement. *Nanomaterials (Basel)*, 2016, **6**(12): p. 1–15.

88. Khotbehsara, M.M., et al. Effect of nano-CuO and fly ash on the properties of self-compacting mortar. *Construction and Building Materials*, 2015, **94**: pp. 758–766.

89. Nazari, A. and S. Riahi. The effects of SiO$_2$ nanoparticles on physical and mechanical properties of high strength compacting concrete. *Composites Part B: Engineering*, 2011, **42**(3): pp. 570–578.

90. Lim, N.H.A.S., et al. Microstructure and strength properties of mortar containing waste ceramic nanoparticles. *Arabian Journal for Science and Engineering*, 2018, **43**(10): pp. 5305–5313.

91. Huseien, G.F., et al. Alkali-activated mortars blended with glass bottle waste nano powder: Environmental benefit and sustainability. *Journal of Cleaner Production*, 2020, **243**: p. 118636.

92. Zhang, A., et al. Comparative study on the effects of nano-SiO$_2$, nano-Fe$_2$O$_3$ and nano-NiO on hydration and microscopic properties of white cement. *Construction and Building Materials*, 2019, **228**: p. 116767.

93. Adak, D., M. Sarkar, and S. Mandal. Structural performance of nano-silica modified fly-ash based geopolymer concrete. *Construction and Building Materials*, 2017, **135**: pp. 430–439.

94. Li, H., et al. Microstructure of cement mortar with nano-particles. *Composites Part B: Engineering*, 2004, **35**(2): pp. 185–189.

95. Praveenkumar, T.R., M.M. Vijayalakshmi, and M.S. Meddah. Strengths and durability performances of blended cement concrete with TiO$_2$ nanoparticles and rice husk ash. *Construction and Building Materials*, 2019, **217**: pp. 343–351.

96. Bodnarova, L. and T. Jarolim. Study the effect of carbon nanoparticles in concrete. In: *IOP Conference Series: Materials Science and Engineering*, 2018. England: IOP Publishing, 385: p. 012006.

97. Madandoust, R., et al. An experimental investigation on the durability of self-compacting mortar containing nano-SiO$_2$, nano-Fe$_2$O$_3$ and nano-CuO. *Construction and Building Materials*, 2015, **86**: pp. 44–50.

98. Khoshakhlagh, A., A. Nazari, and G. Khalaj. Effects of Fe$_2$O$_3$ nanoparticles on water permeability and strength assessments of high strength self-compacting concrete. *Journal of Materials Science & Technology*, 2012, **28**(1): pp. 73–82.

99. Saleh, N.J., R.I. Ibrahim, and A.D. Salman. Characterization of nano-silica prepared from local silica sand and its application in cement mortar using optimization technique. *Advanced Powder Technology*, 2015, **26**(4): pp. 1123–1133.

100. Zhang, S.-L., et al. Effect of a novel hybrid TiO$_2$-graphene composite on enhancing mechanical and durability characteristics of alkali-activated slag mortar. *Construction and Building Materials*, Netherlands, 2021, **275**: 122154.

101. Abbasi, S.M., et al. Microstructure and mechanical properties of a metakaolinite-based geopolymer nanocomposite reinforced with carbon nanotubes. *Ceramics International*, 2016, **42**(14): pp. 15171–15176.

102. Zhang, M.-H. and H. Li. Pore structure and chloride permeability of concrete containing nano-particles for pavement. *Construction and Building Materials*, 2011, **25**(2): pp. 608–616.

103. Mohseni, E., et al. Polypropylene fiber reinforced cement mortars containing rice husk ash and nano-alumina. *Construction and Building Materials*, 2016, **111**: pp. 429–439.

104. Shekari, A.H. and M.S. Razzaghi. Influence of nano particles on durability and mechanical properties of high performance concrete. *Procedia Engineering*, 2011, **14**: pp. 3036–3041.

105. Kaur, M., J. Singh, and M. Kaur. Microstructure and strength development of fly ash-based geopolymer mortar: Role of nano-metakaolin. *Construction and Building Materials*, 2018, **190**: pp. 672–679.
106. Nuaklong, P., V. Sata, and P. Chindaprasirt. Properties of metakaolin-high calcium fly ash geopolymer concrete containing recycled aggregate from crushed concrete specimens. *Construction and Building Materials*, 2018, **161**: pp. 365–373.
107. Gunasekara, C., et al. Effect of nano-silica addition into high volume fly ash–hydrated lime blended concrete. *Construction and Building Materials*, 2020, **253**: p. 119205.
108. Hou, P.-K., et al. Effects of colloidal nanosilica on rheological and mechanical properties of fly ash–cement mortar. *Cement and Concrete Composites*, 2013, **35**(1): pp. 12–22.
109. Shaikh, F., S. Supit, and P. Sarker. A study on the effect of nano silica on compressive strength of high volume fly ash mortars and concretes. *Materials & Design*, 2014, **60**: pp. 433–442.
110. Liu, M., H. Tan, and X. He. Effects of nano-SiO_2 on early strength and microstructure of steam-cured high volume fly ash cement system. *Construction and Building Materials*, 2019, **194**: pp. 350–359.
111. Hosan, A. and F.U.A. Shaikh. Influence of nano silica on compressive strength, durability, and microstructure of high-volume slag and high-volume slag–fly ash blended concretes. *Structural Concrete*, 2020, **22**(S1): pp. 474–487.
112. Revathy, J., P. Gajalakshmi, and M. Aseem Ahmed. Flowable nano SiO_2 based cementitious mortar for ferrocement jacketed column. *Materials Today: Proceedings*, 2020, **22**: pp. 836–842.
113. Potapov, V., et al. Effect of hydrothermal nanosilica on the performances of cement concrete. *Construction and Building Materials*, 2021, **269**: p. 121307.
114. Behzadian, R. and H. Shahrajabian. Experimental study of the effect of nano-silica on the mechanical properties of concrete/PET composites. *KSCE Journal of Civil Engineering*, 2019, **23**(8): pp. 3660–3668.
115. Abd Elrahman, M., et al. Influence of nanosilica on mechanical properties, sorptivity, and microstructure of lightweight concrete. *Materials (Basel)*, 2019, **12**(19): p. 3078.
116. Hameed, M.H., Z.K. Abbas, and A.H.A. Al-Ahmed. Fresh and hardened properties of nano self-compacting concrete with micro and nano silica. In: *IOP Conference Series: Materials Science and Engineering*, 2020. IOP Publishing, 671(1): p. 012079.
117. Andrade Neto, J.D.S., et al. Effect of the combined use of carbon nanotubes (CNT) and metakaolin on the properties of cementitious matrices. *Construction and Building Materials*, 2021, **271**: p. 121903.
118. Zhang, A., et al. Effects of nano-SiO_2 and nano-Al_2O_3 on mechanical and durability properties of cement-based materials: A comparative study. *Journal of Building Engineering*, 2021, **34**: p. 101936.
119. Ma, B., et al. Effects of nano-TiO_2 on the toughness and durability of cement-based material. *Advances in Materials Science and Engineering*, 2015, **2015**: pp. 1–10.
120. Idrees, M., et al. Improvement in compressive strength of styrene-butadiene-rubber (SBR) modified mortars by using powder form and nanoparticles. *Journal of Building Engineering*, 2021, **44**: p. 102651.
121. Wu, Z., et al. Mechanisms underlying the strength enhancement of UHPC modified with nano-SiO_2 and nano-$CaCO_3$. *Cement and Concrete Composites*, 2021, **119**: p. 103992.
122. Alex, A.G., T. Gebrehiwet, and Z. Kemal. M-Sand cement mortar with partial replacement of alpha phase nano alumina. *Journal of Building Pathology and Rehabilitation*, 2021, **6**(1): pp. 1–6.
123. Najafi Kani, E., et al. The effects of nano-Fe_2O_3 on the mechanical, physical and microstructure of cementitious composites. *Construction and Building Materials*, 2021, **266**: p. 121137.

124. Ren, J., Y. Lai, and J. Gao. Exploring the influence of SiO_2 and TiO_2 nanoparticles on the mechanical properties of concrete. *Construction and Building Materials*, 2018, **175**: pp. 277–285.

125. Astm, C., *1202–97. Standard test method for electrical indication of concrete's ability to resist chloride ion penetration*, USA, 1997, 4. pp. 1–8.

126. Lee, H.S., et al. Durability performance of CNT and nanosilica admixed cement mortar. *Construction and Building Materials*, 2018, **159**: pp. 463–472.

127. Zou, B., et al. Effect of ultrasonication energy on engineering properties of carbon nanotube reinforced cement pastes. *Carbon*, 2015, **85**: pp. 212–220.

128. Franzoni, E., B. Pigino, and C. Pistolesi. Ethyl silicate for surface protection of concrete: Performance in comparison with other inorganic surface treatments. *Cement and Concrete Composites*, 2013, **44**: pp. 69–76.

129. Müllauer, W., R.E. Beddoe, and D. Heinz. Sulfate attack expansion mechanisms. *Cement and Concrete Research*, 2013, **52**: pp. 208–215.

130. Scherer, G.W.. Stress from crystallization of salt. *Cement and Concrete Research*, 2004, **34**(9): pp. 1613–1624.

131. Mehta, P.K.. Mechanism of expansion associated with ettringite formation. *Cement and Concrete Research*, 1973, **3**(1): pp. 1–6.

132. Taylor, H.. Sulfate reactions in concrete – Microstructural and chemical aspects. *Ceramic Transactions*, 1994, **40**: p. 61.

133. Algaifi, H.A., M.I. Khan, S. Shahidan, G. Fares, Y.M. Abbas, G.F. Huseien, B.A. Salami and H. Alabduljabbar. Strength and Acid Resistance of Ceramic-Based Self-Compacting Alkali-Activated Concrete: Optimizing and Predicting Assessment. *Materials*, 2021, **14**(20): p.6208.

134. He, X. and X. Shi. Chloride permeability and microstructure of Portland cement mortars incorporating nanomaterials. *Transportation Research Record: Journal of the Transportation Research Board*, 2008, **2070**(1): pp. 13–21.

135. Singh, L.P., et al. Durability studies of nano-engineered fly ash concrete. *Construction and Building Materials*, 2019, **194**: pp. 205–215.

136. Meddah, M.S., et al. Mechanical and microstructural characterization of rice husk ash and Al_2O_3 nanoparticles modified cement concrete. *Construction and Building Materials*, 2020, **255**: p. 119358.

137. Puerto Suárez, J.D., et al. Optimal nanosilica dosage in mortars and concretes subject to mechanical and durability solicitations. *European Journal of Environmental and Civil Engineering*, 2022, **26**(5): pp. 1757–1775.

138. Kumar, S., A. Kumar, and J. Kujur. Influence of nanosilica on mechanical and durability properties of concrete. *Proceedings of the Institution of Civil Engineers – Structures and Buildings*, 2019, **172**(11): pp. 781–788.

139. Scarfato, P., et al. Preparation and evaluation of polymer/clay nanocomposite surface treatments for concrete durability enhancement. *Cement and Concrete Composites*, 2012, **34**(3): pp. 297–305.

140. Çevik, A., et al. Effect of nano-silica on the chemical durability and mechanical performance of fly ash based geopolymer concrete. *Ceramics International*, 2018, **44**(11): pp. 12253–12264.

141. Behfarnia, K. and M. Rostami. Effects of micro and nanoparticles of SiO_2 on the permeability of alkali activated slag concrete. *Construction and Building Materials*, 2017, **131**: pp. 205–213.

142. Saloma, et al. Improvement of concrete durability by nanomaterials. *Procedia Engineering*, 2015, **125**: pp. 608–612.

143. Shaikh, F.U.A. and S.W.M. Supit. Mechanical and durability properties of high volume fly ash (HVFA) concrete containing calcium carbonate ($CaCO_3$) nanoparticles. *Construction and Building Materials*, 2014, **70**: pp. 309–321.

144. Du, S., et al. Nanotechnology in cement-based materials: A review of durability, modeling, and advanced characterization. *Nanomaterials (Basel)*, 2019, **9**(9): p. 1213.
145. Mohammed, M.K., A.R. Dawson, and N.H. Thom. Macro/micro-pore structure characteristics and the chloride penetration of self-compacting concrete incorporating different types of filler and mineral admixture. *Construction and Building Materials*, 2014, **72**: pp. 83–93.
146. Wu, L., et al. Influences of multiple factors on the chloride diffusivity of the interfacial transition zone in concrete composites. *Composites Part B: Engineering*, 2020, **199**: p. 108236.
147. Huseien, G.F., K.W. Shah, and A.R.M. Sam. Sustainability of nanomaterials based self-healing concrete: An all-inclusive insight. *Journal of Building Engineering*, 2019, **23**, pp. 155–171.
148. Hamzah, H.K., et al. Laboratory evaluation of alkali-activated mortars modified with nanosilica from glass bottle wastes. *Materials Today: Proceedings*, 2021, **46**: 2098–2104.
149. Shah, K.W. and G.F. Huseien. Biomimetic self-healing cementitious construction materials for smart buildings. *Biomimetics*, 2020, **5**(4): p. 47.
150. Hamers, R.J.. Nanomaterials and global sustainability. *Accounts of Chemical Research*, 2017, **50**(3): pp. 633–637.
151. Wang, X., et al. Effect and mechanisms of nanomaterials on interface between aggregates and cement mortars. *Construction and Building Materials*, 2020, **240**: p. 117942.
152. Sicat, E., et al. Experimental investigation of the deformational behavior of the interfacial transition zone (ITZ) in concrete during freezing and thawing cycles. *Construction and Building Materials*, 2014, **65**: pp. 122–131.
153. Wang, Y., et al. Beneficial effect of nanomaterials on the interfacial transition zone (ITZ) of non-dispersible underwater concrete. *Construction and Building Materials*, 2021, **293**: p. 123472.
154. Li, L.G. and A.K. Kwan. Concrete mix design based on water film thickness and paste film thickness. *Cement and Concrete Composites*, 2013, **39**: pp. 33–42.
155. Nežerka, V., et al. Impact of silica fume, fly ash, and metakaolin on the thickness and strength of the ITZ in concrete. *Cement and Concrete Composites*, 2019, **103**: pp. 252–262.
156. Dakhil, F.H., et al. Influence of cement composition on the corrosion of reinforcement and sulfate resistance of concrete. *Materials Journal*, 1990, **87**(2): pp. 114–122.
157. Hosan, A., et al. Nano-and micro-scale characterisation of interfacial transition zone (ITZ) of high volume slag and slag-fly ash blended concretes containing nano SiO_2 and nano $CaCO_3$. *Construction and Building Materials*, 2021, **269**: p. 121311.
158. Shoukry, H., et al. Enhanced physical, mechanical and microstructural properties of lightweight vermiculite cement composites modified with nano metakaolin. *Construction and Building Materials*, 2016, **112**: pp. 276–283.
159. Wang, Y., et al. Physical filling effect of aggregate micro fines in cement concrete. *Construction and Building Materials*, 2013, **41**: pp. 812–814.
160. Hong, S.-Y. and F.P. Glasser. Alkali sorption by CSH and CASH gels: Part II. Role of alumina. *Cement and Concrete Research*, 2002, **32**(7): pp. 1101–1111.
161. Hou, P., et al. Effects of mixing sequences of nanosilica on the hydration and hardening properties of cement-based materials. *Construction and Building Materials*, 2020, **263**: p. 120226.
162. Bernard, F. and S. Kamali-Bernard. Numerical study of ITZ contribution on mechanical behavior and diffusivity of mortars. *Computational Materials Science*, 2015, **102**: pp. 250–257.
163. Meng, T., et al. Effect of nano-SiO_2 with different particle size on the hydration kinetics of cement. *Thermochimica Acta*, 2019, **675**: pp. 127–133.

164. Cabrera, J. and M.F.A. Rojas. Mechanism of hydration of the metakaolin–lime–water system. *Cement and Concrete Research*, 2001, **31**(2): pp. 177–182.
165. Snehal, K., B. Das, and S. Kumar. Influence of integration of phase change materials on hydration and microstructure properties of nanosilica admixed cementitious mortar. *Journal of Materials in Civil Engineering*, 2020, **32**(6): p. 04020108.
166. Snehal, K., B. Das, and M. Akanksha. Early age, hydration, mechanical and microstructure properties of nano-silica blended cementitious composites. *Construction and Building Materials*, 2020, **233**: p. 117212.
167. Yaghobian, M. and G. Whittleston. A critical review of carbon nanomaterials applied in cementitious composites – A focus on mechanical properties and dispersion techniques. *Alexandria Engineering Journal*, 2021, **61**(5): pp. 3417–3433.
168. Zheng, Q., et al. Graphene-engineered cementitious composites: Small makes a big impact. *Nanomaterials and Nanotechnology*, 2017, **7**: pp. 1–18.
169. Han, B., et al. Nano-core effect in nano-engineered cementitious composites. *Composites Part A: Applied Science and Manufacturing*, 2017, **95**: pp. 100–109.
170. Lv, S., et al. Effect of graphene oxide nanosheets of microstructure and mechanical properties of cement composites. *Construction and Building Materials*, 2013, **49**: pp. 121–127.
171. Zhao, L., et al. Mechanical behavior and toughening mechanism of polycarboxylate superplasticizer modified graphene oxide reinforced cement composites. *Composites Part B: Engineering*, 2017, **113**: pp. 308–316.
172. Metaxa, Z.S., M.S. Konsta-Gdoutos, and S.P. Shah. Carbon nanofiber-reinforced cement-based materials. *Transportation Research Record*, 2010, **2142**(1): pp. 114–118.
173. Babak, F., et al. Preparation and mechanical properties of graphene oxide: Cement nanocomposites. *The Scientific World Journal*, 2014, **2014**: pp. 1–11.
174. Lu, L. and D. Ouyang. Properties of cement mortar and ultra-high strength concrete incorporating graphene oxide nanosheets. *Nanomaterials*, 2017, **7**(7): p. 187.
175. Peyvandi, A., et al. Effect of the cementitious paste density on the performance efficiency of carbon nanofiber in concrete nanocomposite. *Construction and Building Materials*, 2013, **48**: pp. 265–269.
176. Meng, W. and K.H. Khayat. Mechanical properties of ultra-high-performance concrete enhanced with graphite nanoplatelets and carbon nanofibers. *Composites Part B: Engineering*, 2016, **107**: pp. 113–122.
177. Nazari, A. and S. Riahi. RETRACTED: The effects of zinc dioxide nanoparticles on flexural strength of self-compacting concrete. *Composites Part B: Engineering*, 2011, **42**(2): pp. 167–175.
178. Hogancamp, J. and Z. Grasley. The use of microfine cement to enhance the efficacy of carbon nanofibers with respect to drying shrinkage crack resistance of Portland cement mortars. *Cement and Concrete Composites*, 2017, **83**: pp. 405–414.
179. Ruan, Y., et al. Nanocarbon material-filled cementitious composites for construction applications. In: *Nanocarbon and Its Composites*, 2019. Netherlands: Elsevier: pp. 781–803.
180. Abu Al-Rub, R.K., et al. Mechanical properties of nanocomposite cement incorporating surface-treated and untreated carbon nanotubes and carbon nanofibers. *Journal of Nanomechanics and Micromechanics*, 2012, **2**(1): pp. 1–6.
181. Hilding, J., et al. Dispersion of carbon nanotubes in liquids. *Journal of Dispersion Science and Technology*, 2003, **24**(1): pp. 1–41.
182. Hornyak, G.L., et al. *Fundamentals of Nanotechnology*, 2018. Boca Raton, FL: CRC Press.
183. Nasibulina, L.I., et al. Direct synthesis of carbon nanofibers on cement particles. *Transportation Research Record*, 2010, **2142**(1): pp. 96–101.

184. Makar, J. and J. Beaudoin. Carbon nanotubes and their application in the construction industry. In: *Proceedings of 1st International Symposium on Nanotechnology in Construction, National Research Council Canada*, 2003. Ottawa, Ontario: National Research Council Canada: pp. 331–341.

185. Han, B., et al. Nano carbon material-filled cementitious composites: Fabrication, properties, and application. In: *Innovative Developments of Advanced Multifunctional Nanocomposites in Civil and Structural Engineering*, 2016. Netherlands: Elsevier: pp. 153–181.

186. Yazdanbakhsh, A., et al. Carbon nano filaments in cementitious materials: Some issues on dispersion and interfacial bond. *ACI Special Publication*, 2009, **267**: pp. 21–34.

187. Nasibulin, A.G., et al. A novel cement-based hybrid material. *New Journal of Physics*, 2009, **11**(2): p. 023013.

188. Yazdanbakhsh, A., et al. Distribution of carbon nanofibers and nanotubes in cementitious composites. *Transportation Research Record*, 2010, **2142**(1): pp. 89–95.

189. Habert, G. and C. Ouellet-Plamondon. Recent update on the environmental impact of geopolymers. *RILEM Technical Letters*, 2016, **1**: pp. 17–23.

190. Huseien, G.F., et al. Properties of ceramic tile waste based alkali-activated mortars incorporating GBFS and fly ash. *Construction and Building Materials*, 2019, **214**: pp. 355–368.

191. Huseien, G.F., et al. Utilizing spend garnets as sand replacement in alkali-activated mortars containing fly ash and GBFS. *Construction and Building Materials*, 2019, **225**: pp. 132–145.

192. Habert, G., et al. Environmental impacts and decarbonization strategies in the cement and concrete industries. *Nature Reviews Earth & Environment*, 2020, **1**(11): pp. 559–573.

193. Yildirim, G., M. Sahmaran, and H.U. Ahmed. Influence of hydrated lime addition on the self-healing capability of high-volume fly ash incorporated cementitious composites. *Journal of Materials in Civil Engineering*, 2015, **27**(6): p. 04014187.

194. Guo, X., H. Shi, and W.A. Dick. Compressive strength and microstructural characteristics of class C fly ash geopolymer. *Cement and Concrete Composites*, 2010, **32**(2): pp. 142–147.

195. Mehta, K.P.. Reducing the environmental impact of concrete. *Concrete International*, 2001, **23**(10): pp. 61–66.

196. Shalini, A., G. Gurunarayanan, and S. Sakthivel. Performance of rice husk ash in geopolymer concrete. *International Journal of Innovative Research in Science, Engineering and Technology*, 2016, **2**: pp. 73–77.

197. Han, B., X. Yu, and J. Ou. *Self-Sensing Concrete in Smart Structures*, 2014. London: Butterworth-Heinemann, Elsevier, pp. 232–268.

198. Abdel-Gawwad, H. and S. Abo-El-Enein. A novel method to produce dry geopolymer cement powder. *HBRC Journal*, 2016, **12**(1): pp. 13–24.

199. Davidovits, J.. *Geopolymer Chemistry and Applications*, 5th ed., 2020. Saint-Quentin, France: J. Davidovits. (Issue January 2008.): pp. 1–22.

200. Huseien, G.F., et al. Influence of different curing temperatures and alkali activators on properties of GBFS geopolymer mortars containing fly ash and palm-oil fuel ash. *Construction and Building Materials*, 2016, **125**: pp. 1229–1240.

201. Huseien, G.F., et al. Waste ceramic powder incorporated alkali activated mortars exposed to elevated temperatures: Performance evaluation. *Construction and Building Materials*, 2018, **187**: pp. 307–317.

202. Liew, Y.-M., C.-Y. Heah, and H. Kamarudin. Structure and properties of clay-based geopolymer cements: A review. *Progress in Materials Science*, 2016, **83**: pp. 595–629.

203. Lazaro, A., Q. Yu, and H. Brouwers. Nanotechnologies for sustainable construction. In: *Sustainability of Construction Materials*, 2016. Netherlands: Elsevier. pp. 55–78.

204. Jindal, B.B. and R. Sharma. The effect of nanomaterials on properties of geopolymers derived from industrial by-products: A state-of-the-art review. *Construction and Building Materials*, 2020, **252**: p. 119028.

205. Jiao, D., et al. Rheological behavior of cement paste with nano-Fe_3O_4 under magnetic field: Magneto-rheological responses and conceptual calculations. *Cement and Concrete Composites*, 2021, **120**: p. 104035.

206. Barbhuiya, S., S. Mukherjee, and H. Nikraz. Effects of nano-Al_2O_3 on early-age microstructural properties of cement paste. *Construction and Building Materials*, 2014, **52**: pp. 189–193.

207. Li, Z., et al. Effect of nano-titanium dioxide on mechanical and electrical properties and microstructure of reactive powder concrete. *Materials Research Express*, 2017, **4**(9): p. 095008.

208. Aydın, A.C., V.J. Nasl, and T. Kotan. The synergic influence of nano-silica and carbon nano tube on self-compacting concrete. *Journal of Building Engineering*, 2018, **20**: pp. 467–475.

209. Aly, M., et al. Effect of nano clay particles on mechanical, thermal and physical behaviours of waste-glass cement mortars. *Materials Science and Engineering: A*, 2011, **528**(27): pp. 7991–7998.

210. Lee, G.-C., Y. Kim, and S. Hong. Influence of powder and liquid multi-wall carbon nanotubes on hydration and dispersion of the cementitious composites. *Applied Sciences*, 2020, **10**(21): p. 7948.

211. Assaedi, H., F. Shaikh, and I.M. Low. Effect of nano-clay on mechanical and thermal properties of geopolymer. *Journal of Asian Ceramic Societies*, 2016, **4**(1): pp. 19–28.

212. Hamzah, H.K., et al. Effect of waste glass bottles-derived nanopowder as slag replacement on mortars with alkali activation: Durability characteristics. *Case Studies in Construction Materials*, 2021, **15**: p. e00775.

213. Ahmed, H.U., A.A. Mohammed, and A.S. Mohammed. The role of nanomaterials in geopolymer concrete composites: A state-of-the-art review. *Journal of Building Engineering*, 2022, **49**: p. 104062.

214. Faraj, R.H., et al. Systematic multiscale models to predict the compressive strength of self-compacting concretes modified with nanosilica at different curing ages. *Engineering with Computers*, 2021, **38**(3): pp. 2365–2388.

215. Quercia, G., et al. SCC modification by use of amorphous nano-silica. *Cement and Concrete Composites*, 2014, **45**: pp. 69–81.

216. Quercia, G., G. Hüsken, and H. Brouwers. Water demand of amorphous nano silica and its impact on the workability of cement paste. *Cement and Concrete Research*, 2012, **42**(2): pp. 344–357.

217. Ahmed, H.U., et al. Compressive strength of sustainable geopolymer concrete composites: A state-of-the-art review. *Sustainability*, 2021, **13**(24): p. 13502.

218. Mustakim, S.M., et al. Improvement in fresh, mechanical and microstructural properties of fly ash-blast furnace slag based geopolymer concrete by addition of nano and micro silica. *Silicon*, 2021, **13**(8): pp. 2415–2428.

219. Their, J.M. and M. Özakça. Developing geopolymer concrete by using cold-bonded fly ash aggregate, nano-silica, and steel fiber. *Construction and Building Materials*, 2018, **180**: pp. 12–22.

220. Nuaklong, P., et al. Recycled aggregate high calcium fly ash geopolymer concrete with inclusion of OPC and nano-SiO_2. *Construction and Building Materials*, 2018, **174**: pp. 244–252.

221. Angelin Lincy, G. and R. Velkennedy. Experimental optimization of metakaolin and nanosilica composite for geopolymer concrete paver blocks. *Structural Concrete*, 2021, **22**: pp. E442–E451.

222. Ibrahim, M., et al. Effect of incorporating nano-silica on the strength of natural poz-zolan-based alkali-activated concrete. In: *International Congress on Polymers in Concrete (ICPIC 2018)*, 2018. Springer: pp. 703–709.

223. Janaki, A.M., G. Shafabakhsh, and A. Hassani. Laboratory evaluation of alkali-activated slag concrete pavement containing silica fume and carbon nanotubes. *Engineering Journal*, 2021, **25**(5): pp. 21–31.

224. Kotop, M.A., et al. Engineering properties of geopolymer concrete incorporating hybrid nano-materials. *Ain Shams Engineering Journal*, 2021, **12**(4): pp. 3641–3647.

225. Hamed, N., et al. Effect of nano-clay de-agglomeration on mechanical properties of concrete. *Construction and Building Materials*, 2019, **205**: pp. 245–256.

226. Carriço, A., et al. Durability of multi-walled carbon nanotube reinforced concrete. *Construction and Building Materials*, 2018, **164**: pp. 121–133.

227. Yang, Z., et al. Effect of carbon nanotubes on porosity and mechanical properties of slag-based geopolymer. *Arabian Journal for Science and Engineering*, 2021, **46**(11): pp. 10731–10738.

228. Wu, G., et al. Properties of sol–gel derived scratch-resistant nano-porous silica films by a mixed atmosphere treatment. *Journal of Non-Crystalline Solids*, 2000, **275**(3): pp. 169–174.

229. Bogush, G., M. Tracy, and C. Zukoski IV. Preparation of monodisperse silica particles: Control of size and mass fraction. *Journal of Non-Crystalline Solids*, 1988, **104**(1): pp. 95–106.

230. Sadasivan, S., et al. Preparation and characterization of ultrafine silica. *Colloids and Surfaces A: Physicochemical and Engineering Aspects*, 1998, **132**(1): pp. 45–52.

231. Zhang, M.-H., J. Islam, and S. Peethamparan. Use of nano-silica to increase early strength and reduce setting time of concretes with high volumes of slag. *Cement and Concrete Composites*, 2012, **34**(5): pp. 650–662.

232. Deb, P.S., P.K. Sarker, and S. Barbhuiya. Effects of nano-silica on the strength development of geopolymer cured at room temperature. *Construction and Building Materials*, 2015, **101**: pp. 675–683.

233. Rashad, A.M.. A comprehensive overview about the effect of nano-SiO$_2$ on some properties of traditional cementitious materials and alkali-activated fly ash. *Construction and Building Materials*, 2014, **52**: pp. 437–464.

234. Gao, K., et al. Effects of nano-SiO$_2$ on setting time and compressive strength of alkaliactivated metakaolin-based geopolymer. *The Open Civil Engineering Journal*, 2013, **7**(1): 84–92.

235. Khater, H.M.M.. Physicomechanical properties of nano-silica effect on geopolymer composites. *Journal of Building Materials and Structures*, 2016, **3**(1): pp. 1–14.

236. Assaedi, H., F. Shaikh, and I.M. Low. Influence of mixing methods of nano silica on the microstructural and mechanical properties of flax fabric reinforced geopolymer composites. *Construction and Building Materials*, 2016, **123**: pp. 541–552.

237. Said, S., S. Mikhail, and M. Riad. Recent processes for the production of alumina nano-particles. *Materials Science for Energy Technologies*, 2020, **3**: pp. 344–363.

238. Li, W., et al. Effects of nanoalumina and graphene oxide on early-age hydration and mechanical properties of cement paste. *Journal of Materials in Civil Engineering*, 2017, **29**(9): p. 04017087.

239. Hosseini, P., et al. Effect of nano-particles and aminosilane interaction on the performances of cement-based composites: An experimental study. *Construction and Building Materials*, 2014, **66**: pp. 113–124.

240. Guo, X., W. Hu, and H. Shi. Microstructure and self-solidification/stabilization (S/S) of heavy metals of nano-modified CFA–MSWIFA composite geopolymers. *Construction and Building Materials*, 2014, **56**: pp. 81–86.

241. Shahrajabian, F. and K. Behfarnia. The effects of nano particles on freeze and thaw resistance of alkali-activated slag concrete. *Construction and Building Materials*, 2018, **176**: pp. 172–178.

242. Yang, L., et al. Effects of nano-TiO_2 on strength, shrinkage and microstructure of alkali activated slag pastes. *Cement and Concrete Composites*, 2015, **57**: pp. 1–7.

243. Shafaei, D., et al. Multiscale pore structure analysis of nano titanium dioxide cement mortar composite. *Materials Today Communications*, 2020, **22**: p. 100779.

244. Diamantopoulos, G., et al. The role of titanium dioxide on the hydration of Portland cement: A combined NMR and ultrasonic study. *Molecules*, 2020, **25**(22): p. 5364.

245. Rao, S., P. Silva, and J. De Brito. Experimental study of the mechanical properties and durability of self-compacting mortars with nano materials (SiO_2 and TiO_2). *Construction and Building Materials*, 2015, **96**: pp. 508–517.

246. Chen, L., K. Zheng, and Y. Liu. Geopolymer-supported photocatalytic TiO_2 film: Preparation and characterization. *Construction and Building Materials*, 2017, **151**: pp. 63–70.

247. Sanalkumar, K.U.A. and E.-H. Yang. Self-cleaning performance of nano-TiO_2 modified metakaolin-based geopolymers. *Cement and Concrete Composites*, 2021, **115**: p. 103847.

248. Sastry, K.G.K., P. Sahitya, and A. Ravitheja. Influence of nano TiO_2 on strength and durability properties of geopolymer concrete. *Materials Today: Proceedings*, 2021, **45**: pp. 1017–1025.

249. Maiti, M., et al. Modification of geopolymer with size controlled TiO_2 nanoparticle for enhanced durability and catalytic dye degradation under UV light. *Journal of Cleaner Production*, 2020, **255**: p. 120183.

3 Nano-Enhanced Phase-Change Materials

3.1 INTRODUCTION

Thirty percent of a country's aggregate energy consumption are used by the following building types: residential, institutional, commercial and industrial accounts. Sixty percent of the energy used in a building comprised of heating, ventilation and air-conditioning (HVAC) [1]. Phase-change material (PCM) is a preferred building cooling method when compared with other methods as it compliments green building with efficient energy performance [2].

An effective strategy is phase-change technology, where PCM could enhance the building's thermal mass. This means removing heat from indoors, reducing temperature variations and dispersing heat away from the building with the overall impact of increasing the comfort of the occupants. Studies have discovered that PCMs energy saving ranged from 10% to 30% from air-conditioning consumption within various climate in the United States [3]. During the summer, the energy savings could be up to 30% when PCMs are built on building walls. Microcapsules of PCM application results in reduction of internal temperature of a building by up to 4°C, and in a longer period of time, it stops the temperature from reaching more than 28°C.

PCMs are classified as inorganic, organic and eutectic. Types of inorganic PCMs include metal alloys, metals and hydrated salts, whereas an example of organic PCMs is hydrocarbons-based paraffin wax. There are disadvantages of PCM such as thermal instability, corrosive property, subcooling, low thermal conductivity, leakage and phase segregation [4]. In comparison, organic PCMs are sometimes more suitable due to its non-corrosive properties, immense latent heat capacity, congruent melting and self-nucleation, and chemical inertness and thermal stability [4,5].

Dispersion of a controlled amount of nucleating or dispersant agents is a solution in addressing subcooling and phase segregation issue [4]. Nevertheless, PCM has an inherent low thermal conductivity, denoted by "k". This results in low level of responsiveness during which a thermal change occurs rapidly due to charging/discharging process and its lowered storage capacity. Such issue becomes the centre of attention in research related to thermal energy storage [6]. The k values of hydrocarbon-based PCM range from 0.1 to 0.4 W/mK. Noctadecane is a type of PCM, which possess low solid-state thermal conductivity at 0.35 W/mK. Its liquid state, however, is at 0.149 W/mK [6].

Nanomaterials have undergone fast-paced development leading to the emergence of application strategy of its high level of conductive ultra-small nanosized particles including metal oxides, carbon and metals. These could be used to produce nano-enhanced PCM (nePCM) with significant micro-convection [3] and thermal conductivity [7]. There are ample opportunities for nanomaterials to be applied in

DOI: 10.1201/9781003281504-3

cutting-edge phase-change technology. PCM is generating a high amount of interest in its application of nanometre-scaled thermal conductors through nanofibers, nanoparticles, nanosheets, nanotubes and nanofoams [2].

There are three methods in enhancing PCM's thermal conductivity. The first is to incorporate PCM into porous media such as metallic foams and porous carbon, which has high thermal conductivity. The second method is the dispersion of high thermal conductivity metallic nanostructures or nanoparticles of Cu, Ag or Al to the PCM. The third method is through microencapsulation of the PCM [6]. The thermal conductivity and strength of microcapsules' wall could be increased through nanoparticles that are made of silver [8]. An efficient way of improving PCM additive is copper particles due to its high conductivity and low cost [9].

Three types of elements with thermal conductivity property are being studied [10]. The first is carbon-based nanostructures such as graphene nanoflakes, nanoplatelets, carbon nanotubes (CNTs) and nanofibers. The second is metallic oxide such as TiO_2 and MgO. The third is metals such as aluminium, silver and copper. There is a significant improvement on heat transfer through the use of nanoparticles. The nanoparticles that can be applied to achieve this are carbon that possess various morphologies such as ceramic oxide (CuO, Al_2O_3), metallic nitrides (AIN, SiN), metallic carbides (SiC) and stable metals (gold Au) [3].

Nanomaterials that comprise of metals (Cu, Ag and Al), metal oxides (ZnO) and carbon (single wall SWCNT, graphene nanosheets, active carbon, carbon nanofibers (CNFs), expanded graphite sheets) increase PCM's rate of heat transfer [11]. This chapter will discuss the prominent research being conducted on the development of thermal conductivity through dispersion of three primary PCM nano-enhancers, namely, nanometals, nanocarbons and nano-metal oxides.

3.2   NANO-METAL ENHANCER

It is a common knowledge that metal is efficient at heat conductivity. Silver, in particular, is the optimal conductor of heat and electricity when compared with other metals. Its thermal conduction value is approximately 430 W/mK. The next two metals that are close to silver in terms of thermal conductivities are copper and gold. Gold and silver have two major disadvantages, vulnerable to oxidation and high cost. Therefore, copper, at a significantly lower cost, has the advantages in comparison. Despite this, all the three aforementioned metals have been extensively researched as possible solutions in addressing PCM's thermal conductivities.

Al-Shannaq [6] managed to improve PCM's thermal conductivity "k" by 1168% through the use of nano-thick Ag shells. Specially microencapsulated pure PCM could be used to address leakage issues during its change of state from solid to liquid. However, the microencapsulated shell with poor conductivity value k serves as a barrier to achieving a desirable level of heat transfer and energy storage. A method has been formulated to enhance the PCM's microencapsulated k value that involves using a layer of metallic shell to cover the microcapsules. This is done by activating the surface with dopamine and conducting electroless plating as shown in Figure 3.1.

k value increases to 0.189 from 0.062 W/mK when the diameter of uncoated PCM is increased to 26.9 μm from 2.4. While the diameter is retained at 26.9 μm, there is

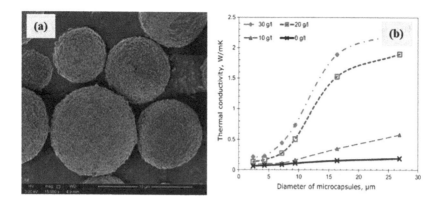

FIGURE 3.1 (a) SEM photo of Ag-coated PCM microcapsules with shell Ag coverage of 70.4 wt%. Silver nitrate of 20 g/L concentration was utilised. (b) Uncoated PCM mean microcapsule diameter and different silver nitrate concentration and its effect on their thermal conductivity [6].

a significant increase of thermal conductivity of metal-coated PCM capsules (2.41 W/mK from 0.189); an increase of 1168%. Such improvement on the thermal conductivity is highly correlated with the size of the shell area that is coated with silver on the surface of the PCM microsphere. The rapid improvement occurs upon the formation of the thermal conduction pathways.

Deng et al. [12] have made another significant improvement (1030%) on the thermal conductivity "k" of the PCM through synthesising AgNWs. First, is the formation of shape-stabilised phase-change materials (polyethylene glycol-silver/EVM ss-CPCMs) composites that occurs after embedding PEG-Ag nanowires into expanded vermiculite EVM. To prevent PCM leakage as well as improving its thermal conductivity, a technique is developed whereby the mixing and embedding are done mechanically. For the purpose of PCM latent energy storage, polyethylene glycol is used. Figure 3.2 shows the silver nanowires that serve as the thermal conductivity promoter. Lastly, the PCM leakage during melting is addressed by having a support matrix (EVM vermiculite) that enhances the mechanical strength.

Significant improvement on the k value of the PEG-infused silver vermiculite composites is achieved through using nanowires with 5–20 μm in length and 50–100 nm in diameter. The result is an increase of 1130% for the k value at 0.68 W/mK when compared to neat PCM with latent heat capacity at 96.4 J/g. The vermiculite has incited supercooling to occur where the temperature has dropped by 7°C upon the PCM for PEG-Ag/EVM ss-CPCMs. Such reaction is similar to nonuniform impregnates for developing nucleation and promoting the formation of PEG crystal. Such improvements are as a result of high k values due to the dispersion of silver nanowire and vermiculite.

Zeng et al. [5] have made an 800% improvement on thermal conductivity using CuNWs. Zeng et al. [5] used CuNWs to enhance thermal conductivity by 800%. Zeng studied the incorporation of copper nanowires (CuNWs) on the k value of

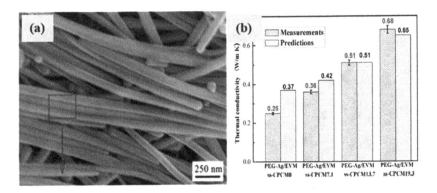

FIGURE 3.2 (a) SEM photos of synthesised silver nanowires. (b) Variation between the predicted thermal conductivity k value and measured values of PCM nanocomposites [12].

tetradecanoyl (TD) as the phase change material. TD with various CuNW weight fractions was synthesized and characterized. The TD was synthesised and classified accordingly to the range of weight fractions of CuNW. The ratio and diameter of free-standing copper nanowires were at 350–450 and 90–120 nm, respectively, with 40–50 μm in length. The CuNW can then be fabricated in bulk through simple technique involving chemical reduction that is water based at room temperature.

Figure 3.3 presents the SEM images that show the composite results, which demonstrate that CuNWs in TD has decent dispersion and entanglement. It is worth noting that the rate of weight loss is lower in comparison to pristine TD due to the structural nature of CuNWs, which is similar to a sponge and is capable of storing the TD within the voids. When the CuNWs is increased by 58.9 wt%, the thermal conductivity increased up to 9-fold, an 800% enhancement.

Zeng et al. [13] managed a 356% increase of thermal conductivity by using AgNWs (380%). Zeng's experiment involved synthesising silver nanowires to produce silver-doped PCM nanocomposites. The inclusion of AgNWs at 45 wt% results in two to three times enhancement of thermal conductivity in graphene-doped PCM. The enthalpy is reduced by 50% and its heat storage capacity has also been reduced. In terms of size, graphene dopants are ten times smaller when compared to doping with silver nanowires. Furthermore, the enthalpy value has also been reduced three times in comparison to AgNWs.

Neat TD's k value was at 0.32 W/mK, and this can be increased to 1.46 W/mK by adding 62.73 wt% nanosilver. As shown in Figure 3.4, the use of AgNWs-TD results in an increase of 356%. In order to create silver nanowires of high quality, PVP (poly(vinylpyrrolidone)) is used with ethylene glycol served as the reducing agent for the silver nitrate. The ethylene glycol performs as a result of "polyol" process.

Shah et al. [14] have increased the PCM thermal conductivity by 160% through the use of copper nanowires (CuNWs). The enhancement of thermal conductivity (more than 50%) of calcium chloride hexahydrate is achieved through adding a trace of CuNWs at 0.17 wt% as shown in Figure 3.5. The use of nanocopper results in

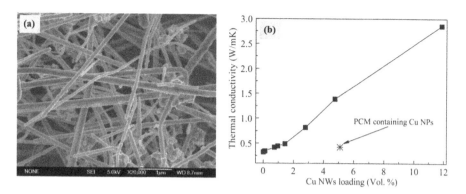

FIGURE 3.3 (a) SEM photos of synthesised CuNWs. (b) Thermal conductivity of PCM composites with increasing CuNWs loadings [5].

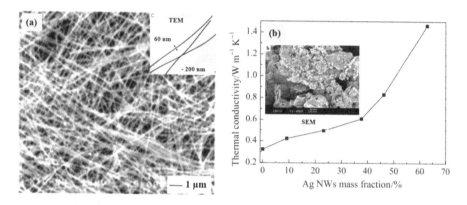

FIGURE 3.4 (a) SEM and (Inset) TEM photos of homogenous self-seeded AgNWs that were fabricated using polyol process. The nanowires were separated from silver nanoparticles through centrifugation. (b) Thermal conductivity enhancement with increasing mass fraction. (Inset) SEM images of silver nanowires Ag NWs-TD composites [13].

optimum enhancement of k value at 160% or an increase to 0.564 W/mK of PCM composite when compared to 0.217 W/mK of neat PCM. Just a trace of CuNWs could result in such significant improvement, thus nanoadditives can be considered as cost-efficient when being applied in buildings.

Metallic nanostructures with length and diameter of 10 µm and 100 nm, respectively, are formed due to controlled and high-yield disproportionation of $CuCl_2$. The reducing agent is oleylamine, which is used within the reaction medium. Unlike conventional method, this does not necessitate the use of reducing chemicals that are harsh. Copper yield is 50% at its highest and is made possible through having control of the initial pre-cursor $CuCl_2$ mixed into reaction medium and reacting temperature. The thermal conductivity of the base PCM that consist of calcium chloride hydrated salt $CaCl_2 \cdot 6H_2O$ can be improved substantially through the use of CuNWs and facile nanosynthesis technique.

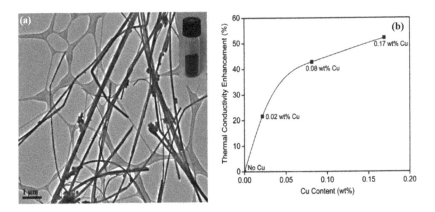

FIGURE 3.5   (a) TEM of nanocopper wires CuNWs. (b) Enhancement of thermal conductivity of PCM calcium chloride salt hydrate $CaCl_2 \cdot 6H_2O$ with respect to CuNWs content [14].

Molefiet et al. [15] have made 70% improvement on the thermal conductivity through the use of copper nanoparticles (CuNPs). The paraffin's thermal conductivity was increased near linearly as the CuNPs were increased. Paraffin wax was used as the PCM base, which was subsequently mixed with molecular-weight polyethylene at low, medium and high rates. The copper particles were mixed with paraffin mixture resulting in the enhancement of the base polyethylene PCM's k value.

Copper particles with a mean size of 30 μm influence the PCM's k value to increase to approximately 70%. The k value is improved due to the copper particles that are in phase-change material polyethylene/wax mixture. It does not have any undesirable effects on other thermophysical properties such as crystalline nature. At the same time, it is able to maintain both thermal stability and mechanical strength.

Tang et al. [16] improved the thermal conductivity "k" value by 38.1% through the use of CuNPs based on SiO_2-embedded-PEG PCM composite that is shape stable. When 2.1 wt% CuNPs were added, the k value was increased by 38.1% when compared to neat PCM. Further addition of copper nanoadditives results in improvement on PEG/SiO_2 hybrid PCMs.

The k value of PEG was 0.297 W/mK, and for PEG/SiO_2, it is 0.36 W/mK. The following are k values (in W/mK) for PEG/SiO_2 composites after they were doped with nanocopper (in wt%): 0.377 W/mK (0.51 wt%), 0.41 W/mK (2.1 wt%), 0.426 W/mK (3.9 wt%) and 0.454 W/mK (6.3 wt%). When PEG/SiO_2 was doped with 2.1 wt% of CuNPs, the thermal conductivity has enhanced by 38%.

Wu et al. [17] have made 30.3% improvement on thermal conductivity through the use of CuNPs. Their results have shown a correlation where 1 wt% of CuNPs could decrease the paraffin PCM heating by 30.3% and cooling by 28.2%. The charging time decreased by 30.3%, while the discharging time was decreased by 28.2% upon the doping of nanocopper particles into the nanocomposites with 1 wt%. Melting PCM heat transfer rate is enhanced through the addition and mixture of nanoadditives (aluminium, copper and copper/carbon nanomaterials). In terms of improvement on heat transfer, nanocopper particles offer the most significant rate amongst others.

The experiment has discovered that the impact on heat transfer is different depending on the nanoparticles used. In terms of raising the temperature up to 67°C, the following nanoparticles and their respective doping measurement were used, copper (1278), aluminium (1311), carbon/copper (1356) and neat paraffin (1416). In terms of decreasing the temperature to 30°C, the following measurement were used: copper (1419), aluminium (1518), carbon/copper (1587) and neat paraffin (1512). The charging time has decreased on the following nanoparticles: copper (9.7%), aluminium (7.4%) and carbon/copper (4.2%). The discharging time has decreased on the following nanoparticles: copper (6.7%), aluminium (0.4%) and carbon/copper (5%). This demonstrates that CuNPs have the prominent role in affecting the heat transfer rate.

Research conducted by Song et al. [8] has discovered that AgNPs are capable of increasing the strength of microcapsule walls. The research has further discovered that the thermal stability is increased by doping AgNPs within the polymer shells of bromohexadecane PCM. The nanoparticles were responsible for the formation of stronger structure thus making it more tolerant to damages due to thermal stresses. AgNPs can be bound tightly with polymeric microcapsules due to its large surface-to-area ratio, enabling it to form a structural composite shell that is strong, and thus the external shell polymer was able to tolerate harsh synthesis process.

Experiments upon thermal gravimetry analysis (TGA) have shown that PCM nanocomposite microcapsules are able to maintain its weight while the temperature rises up to 300°C. This demonstrates that AgNPs doping could significantly increase the polymeric shell structural integrity due to its large surface area and enhance activity and ultra-small-sized nanosilver. The PCM composite is capable of tolerating high temperature in terms of endurance and durability as well as other harsh conditions during industrial production.

3.3 NANO-METAL OXIDE ENHANCER

Two examples of good heat conductors are alumina and copper, both of which are metal oxides with values from 30 to 40 W/mK. Pure metals typically are better heat conductors, but they are not as chemically stable as the metal oxides. In addition, metal oxides are more cost-efficient and reliable in its performance. For these reasons, they are more sought after as a material to replace pure metals.

Babapoor et al. [18] used various nanoparticle types to enhance the thermal conductivity of k value. The following shows the metals with the enhancement percentage: Al_2O_3 (144%), Fe_2O_3 (144%), ZnO (110%) and SiO_2 (110%). Figure 3.6 shows the nanomaterials that were used in the test, which are silica (~20nm), alumina (~20nm), iron oxide (~20nm) and zinc oxide (>50nm). These nanomaterials were added as thermal enhancers and mixed with nanoparticles (SDS) and surfactant (CTAB) for the purpose of synthesising enhanced PCM. The sample that has the highest enhancement at 0.919 W/mK k value was doped with Al_2O_3 nanoparticles.

The doping of nanoparticles gave various enhancement (%) level depending on the concentration (wt%)—Al_2O_3 nanoparticles: 4wt% (120%), 6wt% (141.2%) and 8wt% (144%); Fe_2O_3 nanoparticles: 4wt% (80%), 6wt% (135%) and 8wt% (144%); ZnO nanoparticles: 4wt% (85%), 6wt% (100%) and 8wt% (110%); and SiO_2 nanoparticles: 4wt% (78%), 6wt% (110%) and 8wt% (110%). The results have shown that

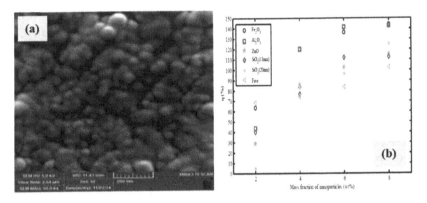

FIGURE 3.6 (a) SEM photo of SiO$_2$ nanoparticles (size ~11 nm) and PCM composite doped with ~8 wt% silica nanoparticles. (b) Enhancement of thermal conductivity with increasing mass fraction of nano-dopants Al$_2$O$_3$, Fe$_2$O$_3$, ZnO and SiO$_2$ [18].

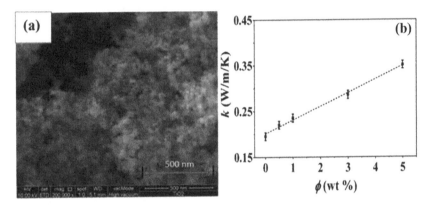

FIGURE 3.7 (a) SEM of TiO$_2$. (b) Thermal conductivity of PA with increasing TiO$_2$ dosages [19].

higher levels of concentration of conductive nanomaterials lead to higher value of k value of the nanocomposites. To conclude, Al$_2$O$_3$ and Fe$_2$O$_3$ have the most significant impacts in terms of enhancing the paraffin-based PCM in itself and its thermal conductivity.

In 2016, Sharma et al. [19] achieved 80% thermal conductivity improvement through the use of TiO$_2$. The study involved the performance of palmitic acid (PA)-based thermal energy storage of synthesised PCM composites that were doped with TiO$_2$ nanoparticles (Figure 3.7). The following is the results of mixing TiO$_2$ into neat PCM with enhancement on k value (%) according to the concentration of TiO$_2$ (wt%): 0.5 wt% (12.7%), 1 wt% (20.6%), 3 wt% (46.6%) and 5 wt% (80%). The high concentration of TiO$_2$ within the PCM resulted in curvilinear characteristic of thermal enhancement.

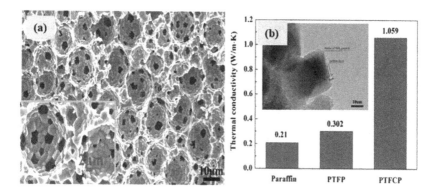

FIGURE 3.8 (a) SEM images of PTF. (b) Enhancement of thermal conductivities of PCM paraffin, PTF/PCM and PTFC/PCM composites by 43.8% and 404%. (Inset) TEM photo of the prepared PCM composite carbonised porous TiO$_2$ foam (PTFC) particles [20].

Li et al. [20] have made 43.8% and 404% improvement on thermal conductivity through the use of TiO$_2$ NPs foam and TiO$_2$ NPs with a nanocarbon shell layer, respectively. The synthesisation of porous TiO$_2$ foam PTFs involved the use of octane as microemulsifier and TiO$_2$ as particle stabiliser (microemulsion technique) as shown in Figure 3.8. The nanosized TiO$_2$ measured at approximately 23 nm consisting of 20% rutile and 80% anatase. Polyacrylic acid-ammonium salt is used as the dispersing agent. It is added on the surface modifier along with a small amphiphilic molecule propyl gallate (C$_{10}$H$_{12}$O$_5$).

The structure of PTFs is porous in addition to continuous connected holes that are 3D in nature. The porosity nature of its structure made it possible to make full absorption of paraffin wax; thus, a surfactant is not required. The structure could also absorb sucrose and would burn off at 1200°C resulting in a thin carbon-based film, which is the carbon nanolayer with a thickness of only 2 nm. Both pure PTF and carbon-based PTF nanocomposites were conductive when compared to pure paraffin with k values of 0.302 and 1.059 W/mK, respectively.

The k value of pure PCM reached 0.302 W/mK when 25 wt% of TiO$_2$ was added. This shows that adding TiO$_2$ enhances the k value, and in this case, it is an increment of 0.092 W/mK. TiO$_2$ foam structure lined with carbon nanofilm with added paraffin exhibits a k value of 1.059 W/m K. This is an increase of 504% when compared with pure paraffin and such significant increase is due to the carbon matrix being adhered onto the TiO$_2$ nanoparticle surfaces. This shows that the novel hybrid nanoparticle TiO$_2$-formed porous foam with inner-lining carbon nanofilms is effective in terms of enhancing PCM for industrial purposes.

Zhang et al. [21] made 18.2% improvement on the thermal conductivity through the use of TiO$_2$. Their research incorporates a novel thermal-insulating film, a polyvinylchloride (PVC) film matrix. Both TiO$_2$ and microencapsulated n-octadecane PCM were used to block UV and act as an additive to regulate temperature. When TiO$_2$ nanoparticles were added at 6 wt%, the k value of the pure micro-PCM reached 0.2356 W/mK from 0.1994 W/mK for the matrix, which is an increase of 18.2%. Such

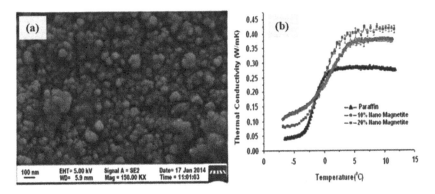

FIGURE 3.9   (a) SEM image of nanomagnetite. (b) Thermal conductivities of PCM paraffin and nanomagnetite-doped PCM (0%, 10% and 20%) [22].

thick film that has excellent heat insulation and thermal regulating properties are suitable to be used for indoor living spaces and cars.

Sahan et al. [22] achieved 60% thermal conductivity improvement by using Fe_3O_4. Figure 3.9 shows the synthesisation and dispersion of iron (III) oxide nanomagnetite composites that were embedded into paraffin PCM. Sol-gel technique, which is very cost-effective was the method of synthesisation of the Fe_3O_4 by using iron chloride hydrates ($FeCl_3 \cdot 6H_2O$, $FeCl_2 \cdot 4H_2O$), hydrochloride and ammonia. They were mixed into paraffin in two concentration levels: 10 and 20 wt%.

Fe_3O_4 nanoparticles have diameter ranging from 40 to 70 nm. Particle aggregation is minimised through surface capping using oleic acid. Uniform dispersion of iron oxide nanoparticles was done on paraffin matrix. The k values have improved by 48% at 10 wt% nanomagnetite mass and 60% at 20 wt%. This shows that PCM that is doped with nanomagnetite particles improves thermal conductivity and it is also cost-effective.

Jiang et al. [23] made 55% improvement on thermal conductivity on PCM through the use of Al_2O_3. The microencapsulation of paraffin (MEPCM) is responsible for the formation of poly(methylmethacrylate-*co*-methylacrylate) polymeric PCM microcapsules. The PCM microcapsules were added with alumina nanoparticles via emulsion polymerisation resulting in significant enhancement. Figure 3.10 shows this significant enhancement where the value has increased to 0.38 W/mK from 0.245 W/mK, which is a 55% increase. There is near parity in terms of the enhancement rate and dosage of nano-Al_2O_3, which means an increase of nano-Al_2O_3 results in higher thermal conductivity of PCM microcapsules.

Tong et al. [24] improved the PCM's thermal conductivity k value through the use of SiO_2. The research used polymeric melamine–urea–formaldehyde for the polymerisation of in situ microcapsules of PCM paraffin before adding graphite and nano-SiO_2. The result of the research revealed that the successful rate of paraffin microencapsulation is at 80% as the PCM paraffin was able to sustain its thermophysical properties. The addition of nano-SiO_2 changed the microcapsules to be resistant to high temperature, reinforcing structural strength of composite and

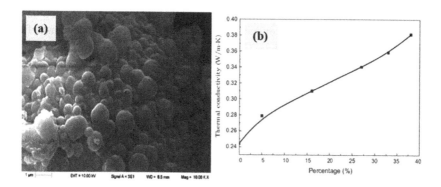

FIGURE 3.10 (a) SEM photos of PCM microcapsules with 27 wt% nano-alumina. (b) Thermal conductivities of PCM with various contents of nano-alumina [23].

high affinity to water. The primary enhancement is of course the k value, which has improved significantly in melting time after the nanomaterials were added.

Ai et al. [25] managed to improve PCM's thermal conductivity through using high-energy planetary milling to develop of ZrO_2 nanopowder-based stearic acid PCM. This study encompasses the heat capability factor (HCF), which is a new concept. Chloroform was used and able to disperse nano-ZrO_2 PCM composites, meaning it is a better alternative for dispersion during ZrO_2 synthesisation in comparison to carbon tetrachloride.

The reason for the better performance using chloroform is due to its capability of improving the surface morphology and sphcrodisation of ZrO_2. The highest HCF is 0.9, which is obtained by having PCM particles size averaging 1.2 μm. However, when the PCM particles size averaging at 0.4 μm, the HCF value reduced to just 0.3. The study thereby concludes that the optimum PCM particles' size is at 1.2 μm to achieve significant enhancements on heat storage capability of chloroform-treated composite ZrO_2-PCM particles.

Song et al. [26] made no enhancement in the attempt to use MgOH NPs for the synthesisation of nePCMs. The supporting materials used were nanosized red phosphorus (RP), MgOH and ethylene propylenediene polymer plastic (EPDM) for the purpose of enhancing the fire resistance attributes in PCM. The fire resistance quality is due to the magnesium hydroxide within the flame retardant shape-stable PCM composite.

The fire-resistant attributes of the PCM could be further improved through reduction of the nanoparticle's diameter. This is because larger surface volume area of MgOH results in rapid break down and high reactivity when being subjected to combustion. This means that the PCM composite will have higher fire resistance quality.

3.4 NANOCARBON ENHANCER

Carbon has higher thermal conductivity when benchmarked against metals and metal oxides. Graphite, graphene and CNT thermal conductivities can be up to five times higher when compared with silver. Research has increasingly focussed on carbon nanomaterials for thermal conductivities due to their continuous decrease of cost of synthesisation.

Ji et al. [27] improved PCM's thermal conductivity by 1700% through the use of ultrathin graphite foams (UGF). The k value was increased by 18 times after adding UGFs into the PCM at approximately 1.2 vol%. Variation of specific heat fusion or melting temperature was not found. Graphite foams consist of ultrathin graphite connected strips. These strips possess higher k value in comparison against metals and carbon foams in their solid state, which means they have better heat response and thermal properties.

Figure 3.11 shows the relationship between UGFs and paraffin. The foam that is interconnected with graphite strips have a thickness of a few 100 nanometres to micrometres. Therefore, phonon scattering at the inter-boundary surfaces of graphite and PCM has no impact on the k value. The basal plane k value multi-layered graphene can be significantly reduced due to interfacial phonon scattering. The improvement on the k value could be estimated and explained through mixtures rule of the graphite foam–paraffin.

Liang et al. [28] have made 1300% improvement on the thermal conductivity through the use of superoleophilic graphene nanosheets mixed with porous nickel (Ni) foam. The synthesisation of polydimethylsiloxane (PDMS-G-NF)-modified graphene-covered nickel foam involved the use of graphene nanosheets layering on the porous Ni foam surface. This results in the formation of graphene-nickel foam G-NF. Further modifications were done on G-NF surface support through the use of siloxane PDMS for the fabrication of shape-stable PCM composite as shown in Figure 3.12.

PA was used to soak micropores within the PCM composite matrix leading to approximately 2.262 W/mK k value with around 59.02 wt% PCM loading. The increment is 14-fold or 1300% when compared to pure PA at 0.162 W/mK. This means that the k value of organic PCM acids can be significantly enhanced by PDMS-modified graphene-nickel porous foams.

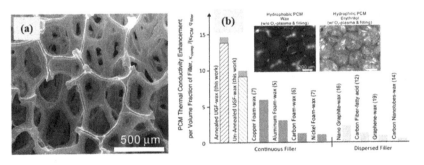

FIGURE 3.11 (a) SEM photo of ultrathin graphite foams without PCM after etching of nickel template. (b) Thermal conductivity enhancement for graphite–paraffin wax and various nanofillers with different volume fractions. All PCMs have k values ranging from 0.17 to 0.31 W/mK. (Inset) Differential interference contrast images of PCM (a) paraffin–graphite foam (b) erythritol–graphite foam [27].

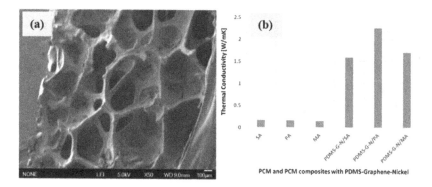

FIGURE 3.12 (a) SEM image of the morphology of micropores of PA-PDMS-G-NF. (b) Thermal conductivity values of *n*-carboxylic acids PCM (SA, PA, MA) and PDMS-graphene-nickel (PDMS-G-NF) foam composites [28].

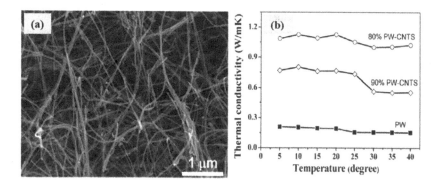

FIGURE 3.13 (a) SEM photo of the interior of CNT porous foam revealing a highly sponge-like microstructure. (b) Thermal conductivities of neat paraffin wax with 10 and 20 wt% loadings of CNT foams, i.e., 80 and 90 wt% paraffin [29].

Chen et al. [29] made 500% improvement on the thermal conductivity "*k*" of PCM through the use of CNT foam. The PCM was absorbed by a permeable support matrix, which is a CNT network with structure similar to sponge. The heat storage capacity of PCM in this research has improved and at the same time can efficiently conduct both heat and electricity. In addition, the PCM composite has the ability to absorb light energy and generate heat via electricity. The PCM composite consists of paraffin-filled soft-flexible CNT-based porous materials as shown in Figure 3.13. The support matrix that is deformable has high rate of thermal conductivity during the solidification and melting processes.

The PCM's thermal properties were significantly influenced by CNT matrix as shown in its high value at 3000 W/mK. The CNT-based sponges composite with 80 wt% paraffin raised the *k* value to 1.2 W/mK or six times more (500%) when compared to pure paraffin wax, which ranges from 0.16 to 0.20 W/mK.

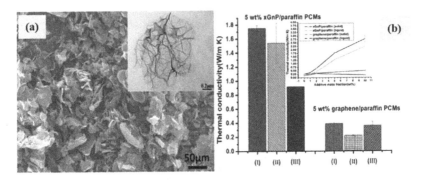

FIGURE 3.14   (a) (Inset) TEM of a single graphene nanoflake and SEM image of exfoliated graphite nanoplatelets (xGnP). (b) Thermal conductivity values of xGnP and graphene-doped PCM using different preparation techniques [30].

Shi et al. [30] used exfoliated graphite nanoplatelets (xGnP) and graphene to increase the thermal conductivity "k" by 1000% and 100%, respectively. Such improvement resulted in the formation of paraffin PCM materials that are stable. Approximately 2 wt% of the graphene was added to paraffin and heated to around 185.2°C. The paraffin retained its form despite reaching significantly high melting point. In order to decrease the cost, trace amount of graphene and xGnP can be doped together in order to improve both stability of form structure and heat dissipation of PCMs as shown in Figure 3.14.

xGnP was doped into PCM resulting in the k value to reach 2.7 W/mK. This is significantly higher when compared to doping using graphene, which returns approximately 0.5 W/mK. It is even more significant when compared to neat paraffin where its k value is at 0.25 W/mK.

Oya et al. [31] used various elements to improve the thermal conductivity "k": graphite (540%), graphite spheres (290%) and nickel spheres (160%). The graphite is exfoliated into PCM erythritol at 118°C, which is its melting point to result in the formation of composite material. Graphite spheres (SG) possess isotropical microstructure attributes, which can be recognised by its multifolds of carbon layerings. The structure of exfoliated graphite (EG) is anisotropical with carbon layers that are flat. Figure 3.15 shows the particles with their diameters, which are SG (8 μm), EG (6 μm) and nickel (5 μm) particles.

The thermal conductivity k value of SG composite (graphite sphere) has increased by 290% at 17 wt% nanofiller loadings when compared to pure PCM erythritol. The k value of expanded graphite is at 4.72 W/mK with 15 wt% nanofiller, which is 540% or 6.4 times more. Nielsen's equation is used for k value calculation and the measurement was conducted using laser flash. Results have shown that expanded graphite possesses higher thermal conductivity when compared to graphite spheres. This is due to many factors including its aspect ratio, larger surface area coverage, percolation clusters and agglomeration packs.

Yavari [32] made 140% improvement on the PCM thermal conductivity through the use of graphene, which was doped into the PCM (1-octadecanol). Thermal

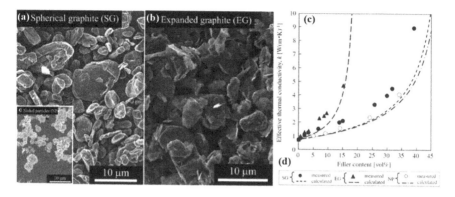

FIGURE 3.15 SEM photos of (a) graphite spheres, (b) expanded graphite EG, and (c) nickel microparticles. Their size dimensions averaged 8, 6 and 5 μm in diameter, respectively. (d) Thermal conductivities of SG (•), EG (Δ) and nickel microparticle (o). The dotted line shows experimental data, and the dashed line shows calculated data [31].

properties on graphene-based PCM were benchmarked against PCM composites using Ag nanowires and MWCNTs (multiwalled CNTs). At 4 wt% nanofiller, the *k* value of PCM composite reached 0.91 W/mK or 140% improvement (2.5 times) along with 15.4% decrease of latent heat capacity in comparison to 0.38 W/mK of neat PCM.

Wang [33] made 305% improvement on PCM thermal conductivity through adding CNFs as nanofillers into the PA. The phase temperature change is approximately 62.5°C after the addition of unwashed acid. The range of length and diameter of the CNFs is 200–500 nm and 5–50 μm, respectively, as shown in Figure 3.16. Alkali potassium hydroxide was used as a treatment for CNFs for the purpose of reducing the thermal boundary resistance of the fibre matrix. Figure 3.16 shows the treated carbon fibres (M-CNF) as well as the surfaced-modified OH functional group into the carbon surfaces.

M-CNF was mixed with PCM with minimal difficulty with the presence of OH (hydroxyl) radicals but without any surfactant. M-CNF in liquid state demonstrates 305.6% improvement on thermal conductivity *k* at 0.5 wt% when compared to 82.6% with 5 wt%. The enhancement for both solid and liquid state is 44.5% and 82.6%, respectively, both with 5 wt% M-CF/PA loading. This means that in liquid state, M-CF has higher thermal conductivity at 0.57 W/mK in comparison to solid state at 0.26 W/mK.

Thermal conductivity of M-CF/PA is 0.2 at 0.5 wt%. M-CF in liquid state, on the other hand, has thermal conductivity at 2.0 at 5.0 wt%. The result has shown that lower loading has optimum thermal conductivity and liquid state has higher enhancement when compared to solid state.

Cui et al. [34] have made improvement on the PCM's thermal conductivity through the use of CNFs by 44% and MWCNT by 24%, both acted as nanofillers. The synthesisation of the composite involved carbon fibres or nanotubes dispersion within both soy wax and paraffin (1, 2, 5, and 10 wt%) at the temperature of 60°C. Figure 3.17

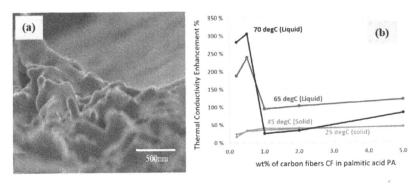

FIGURE 3.16   (a) SEM image of M-CNF/PA with 1.0 wt% M-CNF. (b) Thermal conductivity enhancement with 0.2, 0.5, 1, 2 and 5 wt% of CNF in palmitic acid (PA) [33].

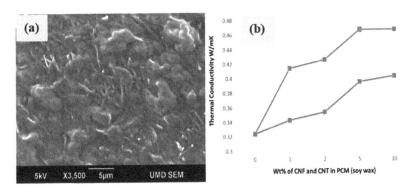

FIGURE 3.17   (a) SEM pictures of CNF/wax composite PCMs. (b) Thermal conductivities of CNF (higher) and MWCNT (lower) in PCM paraffin wax [34].

shows that the increased loading weight of fibres or nanotubes leads to increased k values of the composite PCMs. The enhanced k values demonstrated that nanofibers can significantly enhance the thermal conductivities of paraffin with CNF as additive by up to 44 and 24% using MWCNT. The k value for pure paraffin and PCM composite were 0.320 and 0.450 W/mK, respectively, with a CNF loading of 10 wt%.

Wang et al. [35] made 51% improvement on thermal conductivity through the use of MWCNTs. The MWCNTs was pure and was treated through hydroxyl groups grafting through mechanochemical reaction before being subjected to acid oxidation. PA and PCM composites were doped with CNTs and OH radicals (compared to COOH groups introduced by acidification), respectively. At 1.0 wt%, the k value peaked at 0.339 W/mK or increased by 51.6% when compared with 0.116 W/mK for pure PA.

Wang et al. [36] have improved the PCM's thermal conductivity by 46% through the use of multiwalled CNTs (MWCNT). The synthesisation of MWCNT-PCM composites involved ball milling multiwalled CNTs with potassium hydroxide (KOH).

This method has improved its dispersion in PA. The stability and homogeneity of PCM composites were improved by modifying the grafted OH groups into the MWCNT surfaces. The MWCNT–PA composites with 1 wt% of MWCNT loadings has increased the k values by 46.0% and 38.0% on solid state at 25°C and liquid state at 65°C, respectively.

3.5 SUMMARY

The primary challenge in applying PCM widely as energy solution is that pure PCMs have lower power capacity due to its significantly low thermal conductivity. In the past few years, research has been conducted in improving the thermal conductivity of PCM in addition to retaining high level of energy storage as well as performance related to charging/discharging cycles. This report has explored the various improvements of thermal conductivity through morphologies, nanomaterials and other processes. The conclusions are as follows:

   i. Nanofillers that are continuously interconnected significantly improve thermal conductivity in comparison to nanofillers that are discontinuous and disconnected. The reason behind this is the attributes of matrix foam macrostructures with nanosized walls or struts, which are of higher aspect ratio and more contact surface area that results in higher thermal conductivities. In addition, it also has a higher performance when compared to nanofibers and nanoparticles.
   ii. Both surface tension and the capillary forces between the foam matric and organic PCMs are improved through the oleophilicity effect as a result of surface-modified nanofoams, the overall process in which the PCM thermal conductivities are enhanced. In addition, PCM made with paraffin could stop leakages due to its dispersion property within carbon foams as well as improved retention of PCM within the porous structure.
   iii. Nano-enhancers that are carbon based has higher thermal conductivity when compared to materials that are based on metal or oxide. High surface affinity between the organic structure and carbon nanofillers of PCM enhances uniform interpenetration and decreases phonon scatterings at interfacial surfaces.
   iv. Typically, PCM with carbon-based nanomaterials that is organic in nature has smaller mass density when compared to those that have a higher concentration of metals or metal oxides. Therefore, the dispersion rate and homogeneity within mediums tend to be higher for organic-based PCM composites.
   v. Thermal conductivity of PCM composites is usually higher. This effect is more evident in its liquid state when compared to solid state. While the composite is melting, there is little difficulty in separating nanofibers or nanoparticles group. On the other hand, it is not possible to separate or disperse a solid-state agglomerated cluster that are larger in size and group packings.
   vi. In the foreseeable future, there is a significant role for high performance nano-enhanced phase-change material technology, where it will be

applicable in many areas particularly in thermal storage within the sustainable and renewable energy field. This includes solar energy power generation, industrial heat charging/discharging processes, excess heat management and cooling of electronic devices. The current chapter has explored the latest research and development pertaining to phase-change technologies and remarkable thermal performances of nanomaterials.

REFERENCES

1. Mardiana, A. and S.B. Riffat. Building energy consumption and carbon dioxide emissions: Threat to climate change. *Journal of Earth Science and Climatic Change*, 2015, **S3**: pp. 001–003.
2. Ma, Z., M. Wenye Lin, and I. Sohel. Nano-enhanced phase change materials for improved building performance. *Renewable and Sustainable Energy Reviews*, 2016, **58**: pp. 1256–1268.
3. Hussein, A., N. Payam, M.W. Muhd, J. Fatemeh, R.H. Ben, and A.Z. Sheikh. A review on phase change material (PCM) for sustainable passive cooling in building envelopes. *Renewable and Sustainable Energy Reviews*, 2016, **60**: pp. 1470–1497.
4. Yanio, E.M., G. Andrea, G. Mario, and U. Svetlana. A review on encapsulation techniques for inorganic phase change materials and the influence on their thermophysical properties. *Renewable and Sustainable Energy Reviews*, 2017, **73**: pp. 983–999.
5. Zeng, J., F. Zhu, S. Yu, L. Zhu, Z. Cao, L. Sun, G. Deng, W. Yan, and L. Zhang. Effects of copper nanowires on the properties of an organic phase change material. *Solar Energy Materials and Solar Cells*, 2012, **105**: pp. 174–178.
6. R. Al-Shannaq, J. Kurdi, S. Al-Muhtaseb, and M. Farid. Innovative method of metal coating of microcapsules containing phase change materials. *Solar Energy*, 2016, **129**: pp. 54–64.
7. Kibria, M.A., M.R. Anisur, M.H. Mahfuz, R. Saidur, and I.H.S.C. Metselaar. A review on thermophysical properties of nanoparticle dispersed phase change materials. *Energy Conversion and Management*, 2015, **95**: pp. 69–89.
8. Song, Q., Y. Li, J. Xing, J.Y. Hu, and Y. Marcus. Thermal stability of composite phase change material microcapsules incorporated with silver nano-particles. *Polymer*, 2007, **48**: pp. 3317–3323.
9. Shah, K.W., T. Sreethawong, S. Liu, S. Zhang, L. Tan, and M. Han. Aqueous route to facile, efficient and functional silica-coating of metal nanoparticles at room temperature. *Nanoscale*, 2014, **6**: pp. 11273–11281.
10. Khodadadi, J.M., L. Fan, and H. Babaei. Thermal conductivity enhancement of nanostructure-based colloidal suspensions utilized as phase change materials for thermal energy storage: A review. *Renewable and Sustainable Energy Reviews*, 2013, **24**: pp. 418–444.
11. Raam Dheep, G. and A. Sreekumar. Influence of nanomaterials on properties of latent heat solar thermal energy storage materials – A review. *Energy Conversion and Management*, 2014, **83**: pp. 133–148.
12. Deng, Y., J. Li, T. Qian, W. Guan, Y. Li, and X. Yin. Thermal conductivity enhancement of polyethylene glycol/expanded vermiculite shape-stabilized composite phase change materials with silver nanowire for thermal energy storage. *Chemical Engineering Journal*, 2016, **295**: pp. 427–435.
13. Zeng, J.L., Z. Cao, D.W. Yang, L.X. Sun, and L. Zhang. Thermal conductivity enhancement of Ag nanowires on an organic phase change material. *Journal of Thermal Analysis and Calorimetry*, 2010, **101**: pp. 385–389.

14. Sreethawong, T., K.W. Shah, S.H. Lim, and M.Y. Han. A optimized production of copper nanostructures with high yields for efficient use as thermal conductivity-enhancing PCM dopant. *Journal of Materials Chemistry A*, 2014, **2**: p. 3417.

15. Molefi, J.A., A.S. Luyt, and I. Krupa. Investigation of thermally conducting phase-change materials based on polyethylene/wax blends filled with copper particles. *Journal of Applied Polymer Science*, 2010, **116**(3): pp. 1766–1774.

16. Tang, B., M. Qiu, and S. Zhang. Thermal conductivity enhancement of PEG/SiO$_2$ composite PCM by in situ Cu doping. *Solar Energy Materials and Solar Cells*, 2012, **105**: pp. 242–248.

17. Wu, S., D. Zhu, X. Zhang, and J. Huang. Preparation and melting/freezing characteristics of Cu/paraffin nanofluid as phase-change material (PCM). *Energy Fuels*, 2010, **24**: pp. 1894–1898.

18. Babapoor, A. and G. Karimi. Thermal properties measurement and heat storage analysis of paraffin nanoparticles composites phase change material: Comparison and optimization. *Applied Thermal Engineering*, 2015, **90**: pp. 945–951.

19. Sharma, R.K., P. Ganesan, V.V. Tyagi, H.S.C. Metselaar, and S.C. Sandaran. Thermal properties and heat storage analysis of palmitic acid-TiO$_2$ composite as nano-enhanced organic phase change material (NEOPCM). *Applied Thermal Engineering*, 2016, **99**: pp. 1254–1262.

20. Li, J.Y., Y. Li, W. Deng, X. Guan, and T. Wang. Qian preparation of paraffin/porous TiO$_2$ foams with enhanced thermal conductivity as PCM, by covering the TiO$_2$ surface with a carbon layer. *Applied Energy*, 2016, **171**: pp. 37–45.

21. Zhang, H., Y. Zou, Y. Sun, L. Sun, F. Xu, J. Zhang, and H. Zhou. A novel thermal-insulating film incorporating microencapsulated phase-change materials for temperature regulation and nano-TiO$_2$ for UV-blocking. *Solar Energy Materials and Solar Cells*, 2015, **137**: pp. 210–218.

22. Sahan, N., M. Fois, and H. Paksoy. Improving thermal conductivity phase change materials—A study of paraffin nanomagnetite composites. *Solar Energy Materials and Solar Cells*, 2015, **137**: pp. 61–67.

23. Jiang, X., R. Luo, F. Peng, Y. Fang, T. Akiyama, and S. Wang. Synthesis, characterization and thermal properties of paraffin microcapsules modified with nano-Al$_2$O$_3$. *Applied Energy*, 2015, **137**: pp. 731–737.

24. Tong, X., M. Zhang, L. Lei, H. Zhang, and J. Qiu. Research on the modification of microencapsulated phase change material for thermal energy storage by nano–SiO$_2$ and graphite. *New Chemistry of Material (China)*, 2010, **38**(9): pp. 128–130.

25. Ai, D., L. Su, Z. Gao, C. Deng, and X. Dai. Study of ZrO$_2$ nanopowders based stearic acid phase change materials. *Particuol*, 2010, **8**: pp. 394–397.

26. Song, G., S. Ma, G. Tang, Z. Yin, and X. Wang. Preparation and characterization of flame retardant form-stable phase change materials composed by EPDM, paraffin and nano magnesium hydroxide. *Energy*, 2010, **35**: pp. 2179–2183.

27. Ji, H., D.P. Sellan, M.T. Pettes, X. Kong, J. Ji, L. Shi, and R.S. Ruoff. Enhanced thermal conductivity of phase change materials with ultrathin-graphite foams for thermal energy storage. *Energy and Environmental Science*, 2014, **7**: p. 1185.

28. Liang, W., G. Zhang, H. Sun, P. Chen, Z. Zhu, and A. Li. Graphene–nickel/n-carboxylic acids composites as form-stable phase change materials for thermal energy storage. *Solar Energy Materials and Solar Cells*, 2015, **132**: pp. 425–430.

29. Chen, L., R. Zou, W. Xia, Z. Liu, Y. Shang, J. Zhu, Y. Wang, J. Lin, D. Xia, and A. Cao. Electro and photodriven phase change composites based on wax-infiltrated carbon nanotube sponges. *ACS Nano*, 2012, **6**(12): pp. 10884–10892.

30. Shi, J., M. Ger, Y. Liu, Y. Fan, N. Wen, C. Lin, and N. Pu. Improving the thermal conductivity and shape-stabilization of phase change materials using nanographite additives. *Carbon*, 2013, **51**: pp. 365–372.

31. Oya, T., T. Nomura, M. Tsubota, and N. Okinaka. Thermal conductivity enhancement of erythritol as PCM by using graphite and nickel particles. *Applied Thermal Engineering*, 2013, **61**: pp. 825–828.

32. Yavari, F., H. Fard, K. Pashayi, M. Rafiee, A. Zamiri, R. Ozisik, T. Tasciuc, and N. Ko-Ratkar. Enhanced thermal conductivity in a nanostructured phase change composite due to low concentration graphene additives. *The Journal of Physical Chemistry C*, 2011, **115**(17): pp. 8753–8758.

33. Wang, J., H. Xie, Z. Xin, Y. Li, and C. Yin. Investigation on thermal properties of heat storage composites containing carbon fibers. *Journal of Applied Physics*, 2011, **110**: pp. 094302–094307.

34. Cui, Y., C. Liu, S. Hu, and X. Yu. The experimental exploration of carbon nanofiber and carbon nanotube additives on thermal behavior of phase change materials. *Solar Energy Materials and Solar Cells*, 2011, **95**: pp. 1208–1212.

35. Wang, J., H. Xie, Z. Xin, and Y. Li. Increasing the thermal conductivity of palmitic acid by the addition of carbon nanotubes. *Carbon*, 2010, **48**: pp. 3979–3986.

36. Wang, J., H. Xie, Z. Xin, Y. Li, and L. Chen. Enhancing thermal conductivity of palmitic acid based phase change materials with CNT as fillers. *Solar Energy*, 2010, **84**: pp. 339–344.

4 Preparation and Properties of Nanopolymer Advanced Composites

4.1 INTRODUCTION

The definition of polymer nanocomposites (PNCs) is the combination of more than one material. In addition, the matrix consists of a polymer with the dispersed phase that possess a minimum of one dimension smaller than 100 nm [1]. Many decades of observation have deduced that incorporating nanofillers in small quantities within the polymer resulted in many improvements on its characteristics such as thermal, barrier, mechanical and flame-retardant properties while its processability is unaffected [1,2]. The optimum nanocomposite design necessitates the individual nanoparticles to disperse homogeneously within a matrix polymer.

The main challenge in terms of dispersion state of nanoparticles is to achieve all the possible enhancements of its properties [1,2]. There is a potential for the nanofillers' uniform dispersion to result in significant interfacial area between the nanocomposites' constituents [2]. There are various factors that influence the reinforcing effect mainly polymer matrix properties, type and nature of nanofiller as well as polymer and filler concentration. Other factors focusing on the particle include its size, aspect ratio, orientation and distribution [3]. There have been numerous types of nanoparticles being used to form the nanocomposites with various polymers including clays [3,4], carbon nanotubes [5], graphene [6,7], nanocellulose [8] and halloysite [9].

It is essential to evaluate the nanofiller dispersion within the polymer matrix. This is because there is a strong correlation between the mechanical and thermal properties with the outcome of morphologies. The degree of nanoparticles separation would result in three possible morphological outcomes [10], namely intercalated nanocomposites, conventional composites (also known as microcomposites) and exfoliated nanocomposites (Figure 4.1). In an event where the polymer is not intercalating between the layers of the silicate, the outcome would be separate phases of composite where its properties are within the same range as seen in conventional composites [11].

An intercalated structure encompasses with at least one extended polymer chain, where it intercalates between silicate layers. The outcome is a consistent order of multilayer morphology with polymer and clay layers that are intercalated. Exfoliated structure would result in the event of complete and orderly dispersion of silicate layers within a continuous polymer matrix [10]. Exfoliated nanocomposites have a large

DOI: 10.1201/9781003281504-4

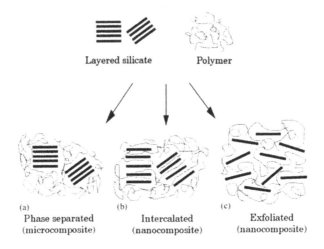

FIGURE 4.1 Possible structures of polymer nanocomposites using layered nanoclays: (a) microcomposite, (b) intercalated nanocomposite and (c) exfoliated nanocomposite [10].

surface contact area between nanoparticles and matrix. Such is one of the significant differences between conventional composites and nanocomposites.

4.2   COMPATIBILISATION IN POLYMER NANOCOMPOSITES

Compatibilisation is of paramount importance to achieve a mixture of polymer or nanocomposite with the desired properties. Therefore, poor properties are attributable to the chemical nature differences between the polymers and polymer matrix with the nanoparticles [12]. As previously mentioned, compatibilisation is a significant factor in obtaining the desired properties. Degradation should be kept at a low probability, and it occurs when the organomodifier is decomposed and when degradation products and polymers are interacting with each other. All of these have a significant influence upon the properties and morphology of the material [13,14] (Figure 4.2).

There are three methods of production for polymer nanocomposites: in situ polymerisation, solution and melt blending. The production method is chosen based on the polymeric matrix type, nanofiller and the final products' desired properties [15].

4.2.1   IN SITU POLYMERISATION

In situ polymerisation involves correct dispersion of the nanofiller within the monomer solution prior to the beginning of polymerisation process. This is to ascertain the formation of the polymer between the nanoparticles. There are various methods of initiating polymerisation such as heat and utilising the correct initiator [16].

This method could be used to achieve a polymer-grafted nanoparticle and high-loading nanofillers with the absence of aggregation [17]. It is possible to include organic modifiers in order to assist the nanoparticles dispersion and to be included within the polymerisation [18]. This method could be deemed as an alternative in

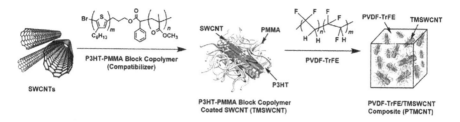

FIGURE 4.2   Scheme of production of compatibilised nanocomposite of PVDF/SWCNT [14].

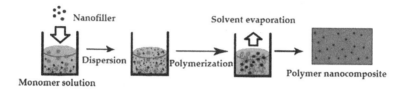

FIGURE 4.3   Schematic illustration for the in situ polymerisation method.

producing nanocomposites through the use of polymers that may be deemed unstable thermally or non-soluble [19].

There are occasions where this method are applicable in solvent-free form [20]. Furthermore, this methods may increase the performance of the products [21]. Mini-emulsion polymerisation is dependent on the monomer droplets being produced, which are subsequently dispersed into a solution within a nanoscale [22]. Figure 4.3 demonstrates the stages of polymer nanocomposites production through using this method.

The advantages include particle morphology that can be controlled [23], high functioning interfacial adhesion of the nanofillers [24] and higher transparency value [25,26]. This method could potentially [15] use higher nanofillers with no presence of agglomeration, increased performance of the final products, products with solvent-free form, outcome of covalent bond within the nanoparticle functional groups and polymer chains as well as utilising the thermoplastic and thermoset polymers. A major disadvantage of such method is the agglomeration easing [17,19].

4.2.2   Solution Blending

Blending is the most used method because it is simple in terms of producing polymer nanocomposites. When compared with other methods, however, this method has higher difficulty in terms of achieving proper nanofiller dispersion within the polymer matrix [15,16].

Solution blending is a system that encompasses both the polymer and nanofiller that can be dispersed within a suitable solvent without much difficulty [16]. The dispersion of the nanofiller within the polymer can be achieved through magnetic stirring, ultrasonic irradiation or shear mixing [17]. Figure 4.4 demonstrates the use of this method, whereby the nanoparticle is still dispersed within the polymer chains after the solvent evaporates. This nanocomposite that has just been produced could be developed into a thin film [15].

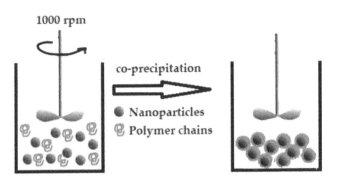

FIGURE 4.4   Schematic illustration for the solution blending method.

The solution blending posed a few constraints in economic and environmental terms. Thus, there is a need for an optimum method to achieve the desired product while addressing the constraints accordingly [27]. Advantages of solution blending include reduced gases permeability [28], simple operation and the use of conventional method for nanofillers of all types, and the thermoset polymers and thermoplastic polymers [29]. The disadvantages include environmental and aggregation issues [27,30]. However, the method is restricted to water-soluble polymers [31].

4.2.3   MELT BLENDING

Melt blending necessitates the direct dispersion of nanofillers into the molten polymer. When the mixing process has started in its melt state, the resulting polymer strain that is applied on the particles depends on the weight distribution and the weight of the molecule. The size of the agglomerates decreases when the shear stress level is high [15]. Figure 4.5 shows the shear flow mechanism during the nanoparticle's dispersion and distribution. At the beginning, the larger agglomerates break apart to become smaller in size before being dispersed within the polymer matrix. Stronger shearing results when the polymer strain is transferred to the new agglomerates. Individual particles are formed due to the breaking down. The primary element of this method is the timing and the chemical processes between the nanoparticles' surface and the polymer [13,32].

Melt blending necessitates single- and twin-screw extruders [33]. However, there are occasions where unfavourable outcome may ensue on the nanofiller's modified surface due to high temperatures, and thus optimisation is applied to address this issue [34]. The most renowned method to address this is the use of intermeshing co-rotating twin-screw extruders. The disadvantage of this method is the difficulty in controlling the parameters such as interaction between the nanoparticles, polymer and the procession conditions such as residence time and temperature [35]. As such, it is not easy to achieve nanoparticles that are evenly dispersed. Figure 4.6 shows the medium dispersive screw profile for a twin-screw extruder. The design encompasses transport, kneading block elements as well as having a turbine element at the melting zone's end [36].

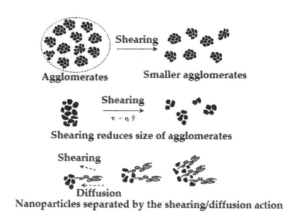

FIGURE 4.5 Effect of shearing on the dispersion of the nanoparticles during melt blending [36].

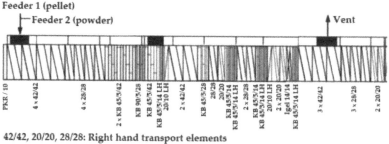

42/42, 20/20, 28/28: Right hand transport elements
20/10 LH: Left hand transport elements
KB 45/5/14, KB 45/5/28, KB 45/5/42, KB 90/5/28: Right hand kneading blocks
KB 45/5/14 LH: Left hand kneading blocks
Igel 14/14: Turbine element

FIGURE 4.6 Schematic illustration of a screw profile of a twin-screw extruder [36].

Melt blending can be commercialised as it is compatible with a range of industrial operations including extrusion and injection moulding [15]. The main advantages of this are low cost, environmentally sustainable due to the absence of solvents, heat stability enhancement [3793], improved mechanical properties [37–45] and good nanoparticles dispersion [46]. Its advantage is the possibility of damage on the nanofillers' modified surface as a result of the high-temperature application [47,48]. Overall, each method has its own respective advantages and disadvantages, and the selection should be based on the conditions and underlying materials.

4.3 NANOPOLYMER FIBRE-REINFORCED COMPOSITES

Gojny et al. investigated the combination of traditionally used fibres such as glass, carbon and aramid fibres along with nanophase reinforcement such as carbon nanotubes [49]. The investigation has revealed 20% increase in matrix-dominated

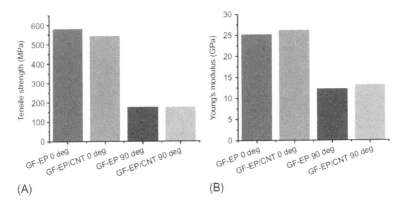

FIGURE 4.7 Experimentally obtained tensile properties of the GFRPs: (a) tensile strengths and (b) Young's modulus in 0 and 90 direction [49].

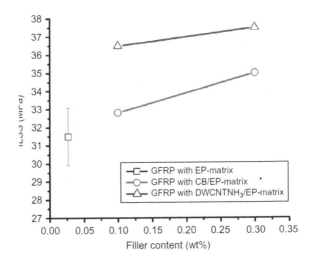

FIGURE 4.8 Interlaminar shear strength (ILSS) of the (nanoreinforced) GFRPs [49].

interlaminar shear strength (ILSS) when a trace amount of carbon nanotubes at 0.3 wt% DWCNT-NH2 is applied. Simultaneously, the CNTs do not affect the tensile properties where it is still fibre-dominated. Figures 4.7 and 4.8 show the experimental derivation of tensile and nanoreinforced composites ILSS, respectively.

Furthermore, there is a higher electric conductivity within the plane compared to an order of higher magnitude in z-direction. Subramaniyan et al. [50] studied the nanoclay dispersion morphology within resin and the acetone can be suspended through the use of transmission electron microscopy (TEM) and scanning electron microscopy (SEM).

In addition, when producing the traditional fibre-reinforced composites with nanoclay, they apply vacuum-assisted wet layup (WAWL). The outcome of this procedure

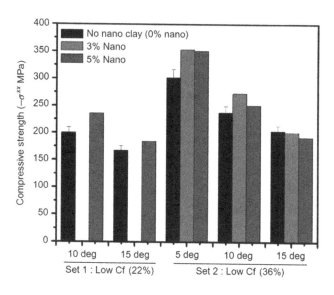

FIGURE 4.9 Compressive strength results—VAWL specimen (% numbers indicate the improvement (+) or reduction (−) in strength due to addition of nanoclay) [50].

leads to improvement upon the nanoclay-enhanced fibre composites' compressive strength [51]. Furthermore, the compressive strength of the fibre-reinforced composites can be predicted by using the elastic–plastic model. The matrix properties were used as the basis for the model and the predicted outcome is very close to the experimental values. They have also discovered a noteworthy phenomenon whereby when the off-axis angles were increased, the compressive strength enhancement is decreased. Specifically, as shown in Figure 4.9, higher volume of fibre in fraction laminates leads to a 5% reduction within the compressive strength at 15 degrees off-axis angle in nanoclay laminates.

4.4 NANOPOLYMER FIBRE-REINFORCED SANDWICH COMPOSITES

Mahfuz et al. [52] have explored the possibility in enhancing the TiO_2 nanoparticle-infused polyurethane foam's performance with regular S-2 glass fibre preforms and SC-15 epoxy sandwich composites. They attempted to increase the core's strength though compromising the toughness of the debond fracture of the sandwich construction. Their results have shown the flexural strength and stiffness has improved substantially along with 3% loading of TiO_2 nanoparticles. Table 4.1 shows Mahfuz et al. study's experimental data.

Debond fracture toughness parameters (Gc) act as the nanoparticle infusion whereby Gc value is decreased close to a factor of three. However, such reduction does not decrease the nanophased sandwich strength where it has risen over the neat system by approximately 53%, Yeh and Hsieh [53], on the other hand, explore the

TABLE 4.1

Experimental Data of Sandwich Composites [52]

Type of Core	No. of Specimens	Maximum Load, *N*	Average Load, *N*
Net polycore	1, 2, 3	730, 596, 774	700
TiO$_2$	1, 2, 3	1170, 1047, 1003	1073

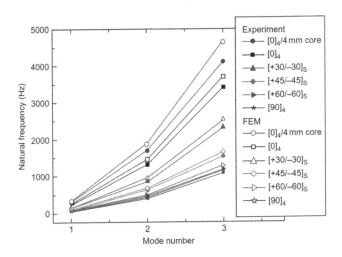

FIGURE 4.10 Natural frequencies of sandwich beams from experiment and finite element analysis [53].

dynamic properties of graphite/epoxy face sheet sandwich beams with MWNT/polymer nanocomposites, which has the potential to be used as core materials.

Their research has discovered that the sandwich beam stiffness is dependent on the face laminate. In addition, the face materials influence the natural frequencies of sandwich beams, where it decreases as the fibre orientations of the graphite/epoxy face laminates increases. The natural frequencies and the sandwich beams' loss factors have increased as the cores' thickness increases. The improved cross-linking has led to a greater loss of phenolic resin when compared to the epoxy resin. The result of the analysis is in accordance with the analysis conducted for finite element. Figure 4.10 demonstrates the sandwich beams' natural frequencies within a variety of cases that were derived from both the experiment and finite element analysis.

4.5 NANOPOLYMERS AND THEIR APPLICATIONS

Amorphous, crystalline and semicrystalline are classifications of polymers. Amorphous polymers have transparent and nonordered properties that are used to produce plastic water bottles. It is easier to produce nonordered amorphous lower as its glass transition temperature is lower when compared to semicrystalline polymers and crystalline polymers. Semicrystalline polymers can be either ordered or nonordered regions depending on the crystalline and amorphous region within the matrices of the polymers.

The arrangement of the molecules in semicrystalline polymers results in higher glass transition temperature properties and stronger when compared to amorphous polymers. When compared to semicrystalline polymers, crystalline polymers possess stronger mechanical properties due to its highly ordered polymers. The reason for this is that crystalline polymer is able to form a resistance against various types of forces due to its crystal lattice formation.

Polymers change their properties when they are synthesised into nanopolymers. The reason for such phenomenon is because macromolecular polymers are significantly larger when compared to nanopolymers. Such properties mean that nanopolymers have their exclusive interaction approach with their environment. An example is the comparison between a nanopolymer and a centimetre-long polymer. The nanopolymer could withstand a shear force when applied.

Nanopolymers have such a wide range of applications that they could be used for all the current polymers' applications. Such applications include telecommunications, defence, household goods, daily services, utilities and basic utilities. Further details include plastic containers, toothpaste and the like. Nanopolymers are favoured due to their chemical properties that are highlighted and sought after such as good chemical resistance, high tensile strength, and capable of holding metals and other compounds.

An example of this is where nanopolymer could be used to create nano circuit due to its high conductive properties. It is possible to produce nanopolymers from many different structures where few can be self-assembled such as lamellar, lamellar-within-cylindrical, lamellar-within-spherical, spherical-within-lamellar and cylindrical-within-lamellar geometry. Other structures that are not self-assembled include polymeric nanocapsules, polymer brushes, nanofibres, hyperbranched polymers, dendrimers and polymeric nanotubes.

4.6 ENVIRONMENTAL APPLICATIONS

There is still constant renovation in nanotechnology with research that is still ongoing pertaining to its usage. Nanofibres are made via electrospinning and have shown to have many possibilities in its usage within the environment. They have remarkable length and able to embed themselves in other media, thus making them one of the safest nanomaterials. Other desirable properties include high porosities at over 80%, adjustable functionality and high surface-to-volume ratio. Such properties are highly effective when compared to the conventional non-woven and polymeric membranes especially pertaining to liquid filtration and particulate separation.

It is feasible for applying nanofibrous scaffolds exclusively and as a cutting-edge component as a separation media for liquid and gas filtration due to advancement in technology including electroblowing and electrospinning. This section will discuss applications of membranes used to remove heavy ion in industrial wastewater and solar energy.

4.6.1 HEAVY ION REMOVAL

Wastewater contains toxic metals that are detrimental to the environment and public health. Removal of such metals is significant in environmental contribution and it is made possible through absorption technology which includes activated charcoal, ion-exchange resins and ion-chelating agents. This method has contributed significantly to the environment when compared to traditional precipitation methods where they are prone to re-pollute the water due to high level of difficulty in removing the toxic metal.

The four major methods of separating technology, involving chemical processes that remove heavy metal from wastewater, are electrolytic recovery, chemical precipitation, solvent extraction/liquid and adsorption/ion exchange. Figure 4.11 shows the chemical separation process and their categorisation of the various treatment types.

The most renowned method of removing ionic metals from a particular solution (e.g., wastewater) is chemical precipitation. Chemical reactions occur between the soluble metal compounds and the precipitating agent. This, in turn, converts ionic metals into insoluble forms before being settled or filtered as part of the removal process. The detailed procedures of such method are precipitation, neutralisation, solids/liquid separation, dewatering and coagulation/flocculation.

Hydroxide precipitation or precipitation by pH is a renowned process for chemical precipitation. This process involves the application of calcium hydroxide (lime) or sodium hydroxide (caustic) to act as the precipitant to form metal hydroxides. Dissolved metal has their own respective pH level (chromium: 7.5 and cadmium: 11.0) for the initiation of hydroxide precipitation. Metal hydroxides are more soluble when the pH values are both low and high. Such property is known as amphoteric, and thus optimum pH for precipitation in a metal may lead to another metal to turn

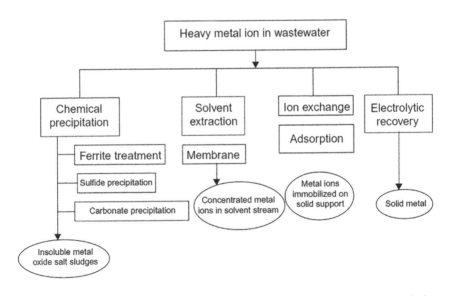

FIGURE 4.11 Various chemical treatment methods for heavy metal removal from wastewater.

into a solution or solubilise. Most process wastewaters contain mixed metals and so precipitating these different metals as hydroxides can be a tricky process.

A popular type of chemical extraction that involves organic solvent is solvent extraction. Typically, it is used in conjunction with various other methods including incineration, solidification/stabilisation or soil washing, which is strictly dependable on the conditions of the site. Despite this, solvent extraction can be used by itself. Both organically bound metals and target organic contaminants could be extracted leaving residuals that necessitates specific handling requirements. The solvent may still remain on the soil matrix that has been treated at trace level. Therefore, the solvent toxicity should be the focus.

Media that has received treatment would typically be returned to the site after being subjected to the most suitable processes along with other required standards. The process proven to be suitable in terms of sediment treatments, sludge and soils that have prevalent organic contaminants including petroleum wastes, VOCs, PCBs and halogenated solvents. The wastes that could be treated using this method include synthetic rubber process wastes, wood-treating wastes, petroleum refinery oily wastes, pesticide/insecticide wastes, paint wastes, coal tar wastes, separation sludge and drilling muds.

Ion exchange is a chemical reaction that is reversible. Firstly, ion is an atom or molecule that has an electrical charge because it has either lost or gained an electron. In terms of wastewater-containing ions, it can be exchanged with another identically charged ion in order to attach to a solid particle that is immobile. Such solid ion-exchange particles could occur naturally through inorganic zeolites or being synthetically manufactured, known as organic resins. Currently, the organic resins that are manufactured are commonly applied as it is possible to control their characteristics during synthesising in order to fulfil the desired applications.

4.6.2 SOLAR ENERGY

Renewable energy is gaining traction in terms of importance due to the deterioration effects of pollutions and limited fossil fuels. One such energy is solar energy where electrical current is produced by using sun's radiant light energy. There has been constant increase on the use of solar energy since the 20th century. This paves the way for homes and business to take advantage of renewable energies for power rather than solely dependent on traditional sources.

There has been constant ongoing research on solar energy. Furthermore, there is also much attention being paid on reducing the cost to store and capture the energy for future use. Currently, solar panels and storage batteries are the most popular form of utilising solar power whereby the power generated via solar panels are stored in the batteries. The stored energy in batteries could then be used in many types of home appliances and machineries. Stored energy is useful in terms of using it during events when radian light is unavailable such as night time or for other reasons.

4.7 POLYMERIC NANOFIBRES AS SENSORS

As the name suggests, nanosensors are sensors that are "nano"-sized, and are measured at 10^{-9} m. Figure 4.12 shows a piece of plastic that has a polymeric nanosensor

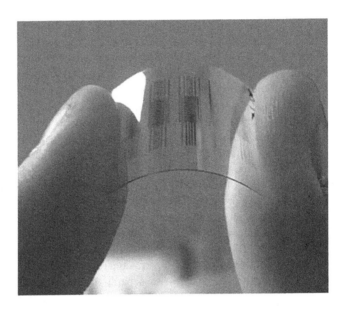

FIGURE 4.12   A nanosensor embedded on a piece of plastic.

attached. The size of the sensor is really notable and as many of them could be attached to the plastic's surface, it can be used to form an "intelligent transportation system". Such system is suitable for traffic and congestion control. The nanosensor's material is carbon nanotube, a composite of plastic or cement. Its application is fascinating because it can be included in pavement due to its cement composite, and thus could be used to monitor traffic flows on roads.

Nanopolymer sensors interact with their environment using electrostatic to cause interactions between its molecules. A specific compound is used to generate an electric current and when this happens, the electrostatic force will channel towards the nanosensors. This, in turn, will push the electrons into the sensors, and thus a current is being forced. The concentration of the compound determines the strength of the current. It can be measured and analysed accordingly. In order to achieve it, calibration is required for the sensor on both ordinary electrostatic force and background.

An example of this is that a sensor should be calibrated for it to recognise and define normal air prior to using it in an environment to measure any hazardous gases. When the hazardous gases reach the appropriate amount, an electrostatic force is exerted into the sensor, where a detection is made the moment the current is formed. Discussion is made pertaining to the fabrication technique of nanopolymers. This section shall focus on the fabrication technique of polymeric nanosensors.

The manufacturing of polymeric nanosensors involves bottom-up assembly and top-down lithography. Bottom-up assembly involves building the nanosensors molecularly. This means assembling the molecules together, meaning positioning each molecule to the desired location to eventually form a larger product. Such production method minimises any defects in the products.

The primary method of producing polymeric nanosensors and other advanced nanopolymers is known as electrospinning, which will be discussed in the chapter. The setback of this level of production is the higher difficulty in achieving mass production of the compound due to its fabrication and assembly of each molecule, which naturally requires a lot of time.

Top-down lithography production breaks down an object that is millimetre or centimetre in size through the use of short-wavelength electromagnetic sources to eventually reach the desired shape for the sensor. The advantage of this form of production is the absence of processes to assemble each of the molecules, which means that mass production is possible. Naturally, products made using this method has higher possibility of defects, naturally because it is taking a larger object rather than assembling them at a molecular level.

The production of polymeric nanosensors involves creating a polymer base and the doping conductive material within the base. The conductive material is typically a semiconductor but other compounds are also possible. There are many processes in doping methods and it is determined by the polymeric base's shape. The primary doping method is known as chemical vapour deposition. Due to a myriad of shapes of the polymeric base, its materials are flexible including nanotubes, nanofibres, nanospheres and other geometrical shapes.

4.8 SUMMARY

To conclude, polymer nanocomposites have vast functionality potentials when compared to the traditional methods. As a result, nanocomposites field has become a popular topic of research due to its many desirable features including production ease, lightweight and flexible. The most distinguishing aspect of polymer nanocomposites is the use of small fillers that result in significant increase in interfacial when compared to the more conventional composites.

Despite the nanofiller's low loadings or differing bulk polymer properties, the interfacial area increases the volume fraction of interfacial polymer significantly. The structure between interfacial and bulk are different, and thus there is a higher surface area in nanoparticles polymer. The interfaces contain the majority parts of the polymer despite the filler's small weight fraction. This is one of the reasons of the difference of reinforcement in nanocomposites. The crucial parameters which determine the effects of fillers on the properties of composites are filler size, shape and aspect ratio and filler-matrix interactions.

REFERENCES

1. Müller, K., et al. Review on the processing and properties of polymer nanocomposites and nanocoatings and their applications in the packaging, automotive and solar energy fields. *Nanomaterials*, 2017, **7**: pp. 74–121.
2. Bitinis, N., M. Hernandez, R. Verdejo, J.M. Kenny, and M.A. Lopez-Manchado. Recent advances in clay/polymer nanocomposites, *Advanced Materials*, 2011, **23**: pp. 5229–5236

3. Zaïri, F., J.M. Gloaguen, M. Naït-Abdelaziz, A. Mesbah, and J.M. Lefebvre. Study of the effect of size and clay structural parameters on the yield and post-yield response of polymer/clay nanocomposites via a multiscale micromechanical modelling. *Acta Materialia*, 2011, **59**: pp. 3851–3863.

4. Oliveira, A.D., N.M. Larocca, D.R. Paul, and L.A. Pessan. Effects of mixing protocol on the performance of nanocomposites based on polyamide 6/acrylonitrile-butadiene-s tyrene blends. *Polymer Engineering and Science*, 2012, **52**: pp. 1909–1919.

5. Maron, G.K., et al. Carbon fiber/epoxy composites: Effect of zinc sulphide coated carbon nanotube on thermal and mechanical properties. *Polymer Bulletin*, 2017, **75**: pp. 1619–1633.

6. de Melo, C.C.N., C.A.G. Beatrice, L.A. Pessan, A.D. de Oliveira, and F.M. Machado. Analysis of nonisothermal crystallization kinetics of graphene oxide-reinforced polyamide 6 nanocomposites. *Thermochimica Acta*, 2018, **667**: pp. 111–121.

7. Gómez, H., M.K. Ram, F. Alvi, P. Villalba, E. Stefanakos, and A. Kumar. Graphene-conducting polymer nanocomposite as novel electrode for supercapacitors. *Journal of Power Sources*, 2011, **196**: pp. 4102–4108.

8. Sonia, A. and K. Priya Dasan. Celluloses microfibers (CMF)/poly (ethylene-co-vinyl acetate) (EVA) composites for food packaging applications: A study based on barrier and biodegradation behavior. *Journal of Food Engineering*, 2013, **118**: pp. 78–89.

9. Marini, J., E. Pollet, L. Averous, and R.E.S. Bretas. Elaboration and properties of novel biobased nanocomposites with halloysite nanotubes and thermoplastic polyurethane from dimerized fatty acids. *Polymer (Guildf)*, 2014, **55**: pp. 5226–5234.

10. Alexandre, M. and P. Dubois. Polymer-layered silicate nanocomposites: Preparation, properties and uses of a new class of materials. *Materials Science & Engineering R: Reports*, 2000, **28**: pp. 1–63.

11. Bhattacharya, S.N., M.R. Kamal, and R.K. Gupta. *Polymeric Nanocomposites Theory and Practice*. Munich/Cincinnati: Carl Hanser Publishers/Hanser Gardner Publications, 2007: pp. 5–10.

12. Mistretta, M.C., M. Ceraulo, F.P. LaMantia, and M. Morreale. Compatibilization of polyethylene/polyamide 6 blend nanocomposite films. *Polymer Composites*, 2015, **36**: pp. 992–998.

13. Mistretta, M.C., M. Morreale, and F.P. La Mantia. Thermomechanical degradation of polyethylene/polyamide 6 blend-clay nanocomposites. *Polymer Degradation and Stability*, 2014, **99**: pp. 61–67.

14. Cho, K.Y., et al. Highly enhanced electromechanical properties of PVDF-TrFE/SWCNT nanocomposites using an efficient polymer compatibilizer. *Composites Science and Technology*, 2018, **157**: pp. 21–29.

15. Mallakpour, S. and M. Naghdi. Polymer/SiO_2 nanocomposites: Production and applications. *Progress in Materials Science*, 2018, **97**, 409–447.

16. Passador, F.R., A. Ruvolo-Filho, and L.A. Pessan. Nanocomposites of polymer matrices and lamellar clays. In: *Nanostructures*, 1st ed., 2017. Elsevier, Netherlands: pp. 187–207.

17. Naz, A., A. Kausar, M. Siddiq, and M.A. Choudhary. Comparative review on structure, properties, fabrication techniques, and relevance of polymer nanocomposites reinforced with carbon nanotube and graphite fillers. *Polymer-Plastics Technology and Engineering*, 2016, **55**: pp. 171–198.

18. Shin, S.-Y.A., L.C. Simon, J.B. Soares, and G. Scholz. Polyethylene–clay hybrid nanocomposites: In situ polymerization using bifunctional organic modifiers. *Polymer (Guildf)*, 2003, **44**: pp. 5317–5321.

19. Lawal, G.I., S.A. Balogun, and E.I. Akpan. Review of green polymer nanocomposites. *Journal of Minerals & Materials Characterization & Engineering*, 2012, **11**: pp. 385–416.

20. Mallakpour, S. and E. Khadem. Recent development in the synthesis of polymer nano-composites based on nano-alumina. *Progress in Polymer Science*, 2015, **51**: pp. 74–93.
21. Modi, V.K., Y. Shrives, C. Sharma, and P.K. Sen. Review on green polymer nanocomposite and their applications. *International Journal of Innovative Research in Science, Engineering and Technology*, 2014, **3**: pp. 17651–17656.
22. Yao, J., Z. Cao, Q. Chen, S. Zhao, Y. Zhang, and D. Qi. Efficient preparation and formation mechanism of polymer/SiO_2 nanocomposite particles in miniemulsions. *Colloid & Polymer Science*, 2017, **295**: pp. 1223–1232.
23. Qi, D., C. Liu, Z. Chen, G. Dong, and Z. Cao. In situ emulsion copolymerization of methyl methacrylate and butyl acrylate in the presence of SiO_2 with various surface coupling densities. *Colloid & Polymer Science*, 2015, **293**: pp. 463–471.
24. Wang, X., L. Wang, Q. Su, and J. Zheng. Use of unmodified SiO_2 as nanofiller to improve mechanical properties of polymer-based nanocomposites. *Composites Science and Technology*, 2013, **89**: pp. 52–60.
25. Yang, F., W. Yang, L. Zhu, Y. Chen, and Z. Ye. Preparation and investigation of water-borne fluorinated polyacrylate/silica nanocomposite coatings. *Progress in Organic Coatings*, 2016, **95**: pp. 1–7.
26. Il, J.C., J. Ko, Z. Yin, Y.-J. Kim, and Y.S. Kim. Solvent-free and highly transparent SiO_2 nanoparticle–polymer composite with an enhanced moisture barrier property. *Industrial and Engineering Chemistry Research*, 2016, **55**: pp. 9433–9439.
27. Kango, S., S. Kalia, A. Celli, J. Njuguna, Y. Habibi, and R. Kumar. Surface modification of inorganic nanoparticles for development of organic–inorganic nanocomposites—A review. *Progress in Polymer Science*, 2013, **38**: pp. 1232–1261.
28. Torabi, Z. and A. Mohammadi Nafchi. The effects of SiO_2 nanoparticles on mechanical and physicochemical properties of potato starch films. *Journal of Chemical Health Risks*, 2013, **3**: pp. 33–42.
29. Zhu, A., H. Diao, Q. Rong, and A. Cai. Preparation and properties of polylactide-silica nanocomposites. *Journal of Applied Polymer Science*, 2010, **116**: pp. 2866–2873.
30. Cong, H., M. Radosz, B.F. Towler, and Y. Shen. Polymer–inorganic nanocomposite membranes for gas separation. *Separation and Purification Technology*, 2007, **55**: pp. 281–291.
31. Bhattacharya, M. Polymer nanocomposites—A comparison between carbon nano-tubes, graphene, and clay as nanofillers. *Materials (Basel)*, 2016, **9**: pp. 262–297.
32. Fornes, T.D., P.J. Yoon, H. Keskkula, and D.R. Paul. Nylon 6 nanocomposites: The effect of matrix molecular weight. *Polymer (Guildf)*, 2001, **42**: pp. 09929–09940.
33. Isobe, H. and K. Kaneko. Porous silica particles prepared from silicon tetrachloride using ultrasonic spray method. *Journal of Colloid and Interface Science*, 1999, **212**: pp. 234–241.
34. Fawaz, J. and V. Mittal. Synthesis of polymer nanocomposites: Review of various techniques. In: *Synthesis Techniques for Polymer Nanocomposites*, 2014. Weinheim, Germany: Wiley-VCH Verlag GmbH & Co. KGaA: pp. 1–30.
35. Fischer, H. Polymer nanocomposites: From fundamental research to specific applications. *Materials Science and Engineering: C*, 2003, **23**: pp. 763–772.
36. Beatrice, C.A.G., M.C. Branciforti, R.M.V. Alves, and R.E.S. Bretas. Rheological, mechanical, optical, and transport properties of blown films of polyamide 6/residual monomer/montmorillonite nanocomposites. *Journal of Applied Polymer Science*, 2010, **116**: pp. 3581–3592.
37. Garcia, M., et al. Polypropylene/SiO_2 nanocomposites with improved mechanical properties. *Reviews on Advanced Materials Science*, 2005, **6**: pp. 169–175.
38. Lin, O.H., H.M. Akil, and Z.A. Mohd Ishak. Surface-activated nanosilica treated with silane coupling agents/polypropylene composites: Mechanical, morphological, and thermal studies. *Polymer Composites*, 2011, **32**: pp. 1568–1583.

39. Grala, M., Z. Bartczak, and A. Różański. Morphology, thermal and mechanical properties of polypropylene/SiO$_2$ nanocomposites obtained by reactive blending. *Journal of Polymer Research*, 2016, **23**: p. 25.

40. Etienne, S., et al. Effects of incorporation of modified silica nanoparticles on the mechanical and thermal properties of PMMA. *Journal of Thermal Analysis and Calorimetry*, 2007, **87**: pp. 101–104.

41. Sun, S., C. Li, L. Zhang, H.L. Du, and J.S. Burnell-Gray. Effects of surface modification of fumed silica on interfacial structures and mechanical properties of poly(vinyl chloride) composites. *European Polymer Journal*, 2006, **42**: pp. 1643–1652.

42. Chau, J.L.H., S.L.-C. Hsu, Y.-M. Chen, C.-C. Yang, and P.C.F. Hsu. A simple route towards polycarbonate–silica nanocomposite. *Advanced Powder Technology*, 2010, **21**: pp. 341–343.

43. Grande, R. and L.A. Pessan. Effects of nanoclay addition on phase morphology and stability of polycarbonate/styrene-acrylonitrile blends. *Applied Clay Science*, 2017, **140**: pp. 112–118.

44. Castro, L.D.C., A.D. Oliveira, M. Kersch, V. Altstädt, and L.A. Pessan. Effect of organoclay incorporation and blending protocol on performance of PA6/ABS nanocomposites compatibilized with SANMA. *Polymer Engineering and Science*, 2017, **57**: pp. 1147–1154.

45. Castro, L.D.C., A.D. Oliveira, M. Kersch, V. Altstädt, and L.A. Pessan. Effects of mixing protocol on morphology and properties of PA6/ABS blends compatibilized with MMA-MA. *Journal of Applied Polymer Science*, 2016, **133**: pp. 1–8.

46. Mallkpour, S. and F. Marefatpour. Novel chiral poly(amide-imide)/surface modified SiO$_2$ nanocomposites based on N-trimellitylimido-l-methionine: Synthesis and a morphological study. *Progress in Organic Coatings*, 2014, **77**: pp. 1271–1276.

47. Dong, Q., et al. Improvement of thermal stability of polypropylene using DOPO-immobilized silica nanoparticles. *Colloid & Polymer Science*, 2012, **290**: pp. 1371–1380.

48. Tanahashi, M. Development of fabrication methods of filler/polymer nanocomposites: With focus on simple melt-compounding-based approach without surface modification of nanofillers. *Materials (Basel)*, 2010, **3**: pp. 1593–1619.

49. Gojny, F.H., M.H.G. Wichmann, B. Fiedler, W. Bauhofer, and K. Schulte. Influence of nanomodification on the mechanical and electrical properties of conventional fibre-reinforced composites. *Composites Part A*, 2005, **36**: pp. 1525–1535.

50. Subramaniyan, A.K. and C.T. Sun. Enhancing compressive strength of unidirectional polymeric composites using nanoclay. *Composites Part A*, 2006, **37**: pp. 2257–2268.

51. Bozkurt, E., M. Kaya, and M. Tanoğlu. Mechanical and thermal behavior of non-crimp glass fiber reinforced layered clay/epoxy nanocomposites. *Composites Science and Technology*, 2007, **67**: pp. 3394–3403.

52. Mahfuz, H., M.S. Islam, V.K. Rangari, M.C. Saha, and S. Jeelani. Response of sandwich composites with nanophased cores under flexural loading. *Composites Part B*, 2004, **35**: pp. 543–550.

53. Yeh, M.-K. and T.-H. Hsieh. Dynamic properties of sandwich beams with MWNT/polymer nanocomposites as core materials. *Composites Science and Technology*, 2008, **68**: pp. 2930–2936.

5 Nanotechnology-Based Smart Glass Materials

5.1 INTRODUCTION

The conventional usage of high-performance glazing systems is upon windows or building windows in order to decrease the amount of unwanted heat from the sun as well as reducing the workload to cool air from the air-condition systems installed within the building. Aesthetically, glass facades are more desirable in terms of using them in commercial buildings [1,2]. This means that research is needed to estimate and evaluate the energy savings in practical terms for the high-performing glass.

It is particularly vital for the research to be conducted on the different types of high-performance building glass. This is especially true for densely built cities such as Singapore, where the heat from the sun is an issue for the buildings. Additionally, it is still uncertain whether these glasses are able to retain its efficiency in countries with four seasons. The glass's U-value needs to be within acceptable range in order for it to be functional when subjected to various climates including the tropics.

Assessment of glass performance necessitates active measurement while being subjected to a controlled source of radiant. Evidently, such testing environment may not account for actual weather, where many possibilities may not be duplicated within the test environment [2,3]. This may be compensated by subjecting the glasses to real weather conditions by actually installing them outdoor [4,5]. However, it is still not possible to conduct testing in large-scare and fast on-site characterisation. Furthermore, the test is restricted to fabricated glazing, which means that it is not possible to forecast possible problems during the design stage.

The development of cutting-edge technique made it possible for professionals to be able to make simulation and assessment on the glass being installed in building during the design stage [4,6,7]. Over-the-counter and matured simulation codes that are open source in nature including Energy Plus and Radiance [4] are appropriate for such assessment as they have undergone development spanning more than 10 years. Despite such tools, the assessment is still complex in terms of conducting such evaluation for high-performance glazing description (glass with various coatings for many purposes) into the solar irradiance module using the current glass models.

Therefore, the glass involving multiple layer/pane glazing must be classified and computed uniquely using more focussed tool prior to be interfaced using a custom script. Further consideration is vital in terms of acquiring complete and comprehensive weather model to increase the accuracy of the solar heat gain that will enter into the building. With the exception of weather data from International Weather for Energy Calculations (IWEC), other data must be carefully considered when inputted for assessment using the aforementioned tools. In addition, the tools are developed with main consideration to indoor performance; thus, it does not account for the

DOI: 10.1201/9781003281504-5

negative effects and other impacts of the sunlight being reflected off the building glass facade during the assessment for environmental risks. External use of the glass typically involves alternative glass material, which are mirror or opaque with high specularity. The use of such alternative materials means that their properties have no angle dependence [4].

The evaluation of glass performance is often conducted using active measurement, which makes use of known radiant source. However, this type of setup cannot be applied for testing under non-controlled weather. Though a passive test procedure can be conducted under actual weather conditions using the outdoor test chamber, it is not suitable for a large-scale testing and fast on-site characterisation. In addition, this test is limited to the fabricated glazing and thus could not predict potential issues in the design stage. Advances in simulation techniques have enabled building professionals to evaluate the glass facade of a building at the design phase. However, the typical simulation tools are unable to integrate the high-performance glazing description, which is generated thanks to advances in coating technology, using the existing glass models. Furthermore, these tools often lack local weather models that play an important role in accurately assessing the solar heat gain admitted into the building.

5.2   NANOTECHNOLOGY

Nanotechnology is being applied in various disciplines especially within construction materials due to its ability to decrease the consumption of energy; thus, they have much potentials. Glass is one particular material for the construction materials and if it can be treated with nanotechnology, it can decrease the transfer of heat through the building envelope as shown in Figure 5.1. The study used Design

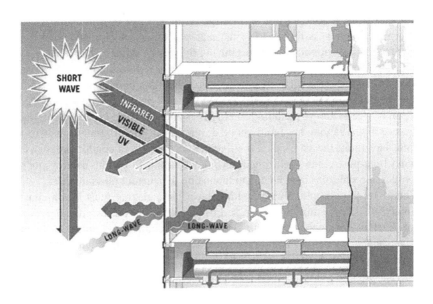

FIGURE 5.1   The glass treated with nanotechnology [8].

Builder 3.1 and followed the Egyptian energy code requirement to assess the difference in energy consumption between two types of glass: standard 6 mm clear glass and glass that is treated with nanotechnology. The standard 6 mm clear glass that was used in glazed facades resulted in passing high thermal loads into the indoor environment of the building. This results in increasing use of energy in the building [8].

At 1 billionth of a metre, nanometre is extremely small. To put it into perspective, it is one hundred thousandth of the width of a strand of hair or length of ten hydrogen atoms (a red blood cell is 2000 nm). Nanomaterial is classified as a premium category of advanced materials. This means that they can be produced as small as 1 or 100 nm for its inner grains when compared to the more conventional material where their dimension is at least 100 nm.

This means that nanomaterials have remarkable properties that were not possible or does not exist in the more conventional materials as shown in Figure 5.2 [8]. It is now recognised that nanomaterials are building materials of the 21st century, where it could be used in the advanced fields such as biotechnology, nanotechnology, information and communication technology. They are more or less the ingredients of progress and the advancement of civilisation across nations as well as being the representative for a renaissance. Further advantages of nanomaterials are its range of varieties and proportions from organic to inorganic, or synthetic to natural materials.

Nanotechnology has established itself in a variety of fields such as electronics, medicine, energy, building and construction, textiles, food, automobile and cosmetics. Specific applications on the aforementioned fields include nanosensors and nanoelectron mechanical systems (NEMS), drug delivery targeting cancer cells, sunscreens, face cream with nanoparticles, water or stain-proof cloth as well as self-cleaning windows for vehicles and within the construction sector.

Glass is a common material within various industries including transport, building and construction. It is also being used in microscopes, tablet computers, furniture and many more. There are four advantages of using glass within the building and construction sectors. Firstly, it allows natural lights to enter the building. Secondly, it filters out harmful rays from the sun from entering the building. Thirdly, it harmonises the environment and the building. Lastly, it is cost-effective due to its energy efficiency. Such benefits of glass made it very suitable in the construction industry and the industry is using it in hard-to-reach areas.

Researchers and scientists have taken the motto of "necessity is the mother of invention". Thus, during their research in improving the properties of glass, they have developed a type of glass that requires minimal maintenance, also known as self-cleaning glass. Individuals that wear glasses will be glad that such glasses prevent mist from forming when they are enjoying hot drinks that are steaming or when they are cooking. In addition, anti-fogging glass is used in tablet computers; thus, they could be used in close proximity to swimming pools. In addition, anti-reflective glass is used in mobile phones or laptops thus could still be able to use these devices during broad daylight. As for self-cleaning glass, they are most suitable for windows and doors in offices and homes, where they don't need to clean them frequently (SCG)

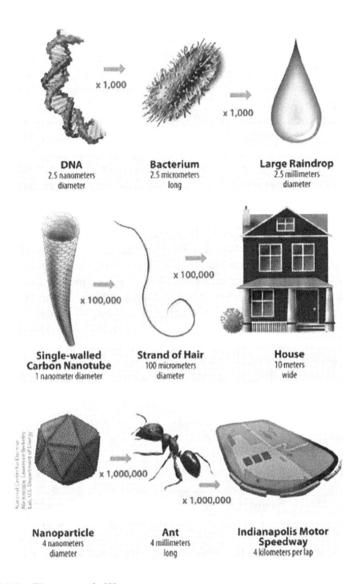

FIGURE 5.2   The nanoscale [8].

5.3   SELF-CLEANING GLASS

Glass is used extensively in many industries such as automotive industry, solar energy cells and building and construction industry. SCG is a new type of glass being developed and is widely used in hard-to-reach areas in buildings because it requires minimal maintenance. SCG either has a layer of titania that measures 10–25 nm or is coated with silica on its surface via both bottom-up and top-down approaches. The self-cleaning properties are the control of its wettability properties on its surface.

The first is for the surface that is completely dry, also known as hydrophobic surface, where a liquid droplet maintains a spherical shape on the surface of the glass. This is achieved either by forming a component of low surface energy or through surface roughness control. The surface becomes hydrophobic by applying a thin layer of SiO_2. The second involves complete wetting of the surface where the liquid forms a film upon contact with the surface. This is known as hydrophilic and can be achieved through applying photocatalytic TiO_2 coating. The coating uses the sunlight and water to rinse itself, thus resulting in self-cleaning property.

Therefore, solid surfaces will result in various reactions with dissimilar materials depending on the type of coating being applied. Consideration should take place upon the various qualities as a result such as spreading, wettability, adhesion and interface. The wettability property of a solid is defined by observing that contact angle (denoted by θ) the moment liquid touches the surface of the solid. Figure 5.3 shows the balance of interfacial tensions that are used to calculate the contact angle.

where σsv, σsl and σlv are the corresponding interfacial energies between the solid–vapour, solid–liquid and liquid–vapour phases, respectively; θ denotes the contact angle defined by the tangent on a liquid droplet with respect to the solid surface; the condition $\theta \leq 90°$ indicates that the solid is wet by the liquid; and $\theta \geq 90°$ indicates non-wetting, with the limits $\theta = 0°$ and $\theta = 180°$ defining complete wetting and complete non-wetting, respectively. Depending upon the contact angle, coatings are of two types: hydrophilic and hydrophobic.

5.4 HYDROPHILIC COATING

A surface is deemed hydrophilic when the water contact angle (CA) is less than 90°. It is considered as super hydrophilic when its CA is less than 50°. As the liquid contacts such surfaces, it will spread out until it becomes a thin layer. The self-cleaning materials that made this possible are WO_3, ZnO, SnO_2, SiO_2, CdS, TiO_2 and ZrO_2. The most extensively used is TiO_2 because it has more advantages when compared with the others.

New discovery made by Fujishima and Honda utilised TiO_2 for photo-electrochemical splitting of water to hydrogen and oxygen while being subjected to UV radiation. This has resulted in an explosion of research to study the TiO_2 photo-catalysis potentials including self-cleaning coatings, photocatalysis, photovoltaics, photoelectrocatalytic degradation of organic compounds and advanced oxidation.

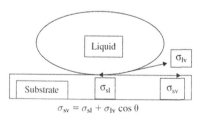

$$\sigma_{sv} = \sigma_{sl} + \sigma_{lv} \cos \theta$$

FIGURE 5.3 Interfacial tensions on a solid substrate.

TiO_2 exhibits the following properties: high refractive index, good mechanical performance, transparent and semiconductor material with a high band gap.

When TiO_2 is within the wavelength range from 0.35 to 12 mm, it becomes stable chemically. Titania has three different crystalline formation: brookite, anatase and rutile. The highest refractive index is rutile at 2.61–2.90 and thus making it the centre of focus for optical applications. When being subjected to ranges of temperatures, rutile is also the most stable in terms of its thermodynamic properties. Despite the advantages, anatase has increased desire for lower temperature applications where it is necessary to form a film on thermally sensitive substrates. Therefore, the desirable materials are either amorphous or crystalline anatase, which are used to produce self-cleaning glasses at temperatures below 400°C. Anatase will change to rutile as it reaches the temperature from 700°C to 1100°C.

5.5 ANTI-REFLECTIVE COATING

Fujishima and Guiselin et al. invented the TiO_2 thin films, and they have also patented the methods. This film is transparent, photocatalytically efficient and abrasion-resistant, which can be used on glass surfaces. Several SCG are already being commercially used at present such as Hydrotecht from TOTO, Activt from Pilkington Glass, Thermotecht from Viridian and Bioclean from Saint-Gobain. Additionally, self-glazing products are also being rolled out in liquid forms or white that target direct consumers.

When a normal glass is applied to the self-cleaning products, they would turn into SCG. Products that are available for users are produced by some companies such as Rain Racert from Rain Racer Developments, BalcoNanot from Balcony Systems Solutions and ClearShieldt from Ritec International. SCG can be installed in various locations including offices, facades and general buildings. The improvement of photocatalytic activity and anatase coating necessitates a high refractive index due to low-temperature processing.

Other properties apart from self-cleaning are necessary for a glass that will be used on smart phones, spectacles and solar cells. These properties include anti-fogging, anti-abrasive and anti-reflection. Fraunhofer is the founder of anti-reflective (AR) coating in 1817. Since then, AR phenomenon is regarded as a destructive interference between air-coating interfaces by Fresnel and Poisson and light reflected due to substrate coating. One of the many methods of making AR coating is to construct a single-layer of coating that has low refractive index. Materials that have low refractive index cost more and also rare.

Porous nanostructures can be used to effectively decrease the volume-averaged refractive indices of materials through controlling the porosity within the coatings. This results in AR coatings, and its hydrophobic and hydrophilic properties could be further enhanced by increasing surface roughness. At the same time, reflection is increased due to the decrease in transmittance, which occurred as a result of scattering diffusion in rough surfaces. Sample transmittance is the subsequent light intensity ratio that exits after the intensity ratio entered the sample.

Therefore, the increase of transmittance results in decrease of photocatalytic activity as the light intensity decreases. At the same time, AR surfaces are part of the

SCG. Therefore, in order to preserve the self-cleaning and anti-reflectivity properties, the ideal surface roughness is required. The assessment of the solid surface's wettability necessitates the static CA and the dynamic sliding angle. The essential factor is therefore roughness of the surface and chemical functionalisation.

5.6 PHOTOCATALYTIC ACTIVITY OF TIO$_2$

When TiO$_2$ is subjected to ultraviolet (UV) light, it will absorb them and in turn forms charge carriers (hole and electron) that is photo-generated, which is consistent with the band gap as demonstrated in Figure 5.4. The TiO$_2$ surface will be diffused with the photo-generated holes in the valence band. A reaction will occur due to the adsorbed water molecules resulting in the formation of hydroxyl radicals (OH$_2$) as presented in Figure 5.5. The organic molecules on the TiO$_2$ that were in close approximation to photo-generated holes and the hydroxyl radicals were oxidised.

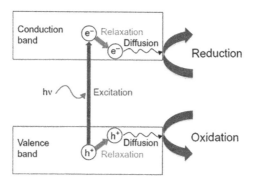

FIGURE 5.4 Schematic illustration of the formation of photo-generated charge carriers (hole and electron) upon absorption of ultraviolet (UV) light.

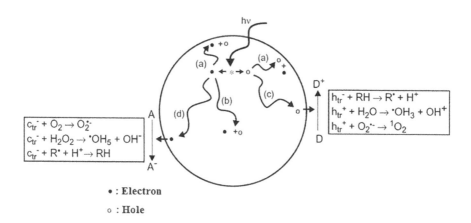

FIGURE 5.5 Reactions occurring on bare TiO$_2$ after UV excitation.

The reduction processes occur due to the electrons located in the conduction band, after which it will react with the natural oxygen molecules, thus forming superoxide radical anions. This photocatalysis reaction is the key factor of the self-cleaning properties of the glasses.

5.7 FABRICATION OF SELF-CLEANING GLASS

A glass becomes either hydrophilic or hydrophobic after the applications of a thin layer of TiO_2 or SiO_2 on its surface. There are two types of fabrication of nanomaterials which are top-down and bottom-up. The top-down approach involves removal of materials gradually from massive structure until the required nanomaterial is formed. Lithography is an example of this method. Comparatively speaking, it is similar to using a block of wood and turning it into a doll by a carpenter.

Bottom-up approach involves the use of atoms or molecules to be built gradually until the formation of the required nanomaterial or nanocoating. Comparatively, this is akin to using Lego blocks to build a house. The bottom-up approach is further divided into two sub-methods, which are gas and liquid phase. The gas-phase method involves the evaporation of plasma arc and chemical vapour deposition (CVD). The liquid phase involves the sol-gel and molecular self-assembly.

5.8 SIO₂–TIO₂ COATING

In addition to having self-cleaning function, other functions are also desirable including photocatalysis and anti-reflectivity, which are vital in products such as smart phones and solar cells. Glop et al. [9] used the sol-gel method, and Liu et al. [10] used the pulse magnetron sputtering as methods of preparation for the photoactive anti-reflection coating. The TiO_2 coating on the outer surface that results in self-cleaning feature increases the reflectance of plastic or glass substrate due to its relatively high refractive index (c. 2.5 for the anatase phase).

Therefore, self-cleaning and anti-reflectivity attributes may not be compatible with the exception of rare instance where the structure and composition are modulated. Prado et al. [11] attempted to produce a coating that is multifunctional where its outer layer consists of dense/mesoporous TiO_2 and its inner layer consists of mesoporous SiO_2 AR layer. Multifunction coatings with self-cleaning attribute were discovered to perform 25%–30% when compared to photodegradation degree, which is produced via the conventional TiO_2 coatings layer that are either porous or compact.

Solar industry including solar power plants and solar energy producers primarily use glass. The amount of electricity generated or power for heating water depends on the intensity of the sunlight. The glass may be useful in terms of reducing loss of radiation and reflection.

SCG's primary property is its reflective index denoted by n. Producing glasses with AR properties for solar-related use requires a low refractive index such as SiO_2 where its n value is 1.4. Conversely, high reflective index such as titania where its n value is 2.0 is vital in improving the photocatalytic activity of the hydrophilic property in SCG. Helsch and Deubener [12] attempted to use the sol-gel coating technique to create a single type of glass that contains both functions. High transmittance is needed for this particular type of glass. The research has been a success where they used two layers consisting of SiO_2/TiO_2 to create a glass with both AR and photo-catalytic properties.

Through the sol-gel coating method, preparation was made on silica glass porous coatings $xTiO_2$ (100-x) SiO_2 with 50 wt% titania. The compatibility of AR and photocatalytic properties will then be achieved once the composition reached the ranger x from 7.5 to 20. Porous coating also enhances the solar transmittance by 2.3% when compared to silica glass that is not coated. These coatings with dual functions have greater degradation rate at 20-fold of the airborne contaminants when compared to nanoporous film of pure SiO_2.

Nanoporous structures consist of materials that have high porosity and low density, and have simultaneous advantages in terms of possessing high pore volume, high surface area and larger pore size, whereby the diffusion pathways are accessible. Anti-reflection coating is often made of porous silica layers. Helsch et al. [13] made a discovery that there is a 5% enhancement (from 92% to 97%) of light transmission when borosilicate glass is at 550 nm, 35% porosity and 110 nm film thickness. There are many applications for TiO_2 films on glass substrates including mirrors, windshields and window glasses.

While being serviced, AR porous coatings will be subjected to severe environmental conditions including hail and salt atmosphere, sandstorms, dust particles and airborne volatile organic compounds. If the AR coating is damaged while being subjected to the aforementioned conditions, the solar transmittance will be reduced. Cathro et al. [14] discovered an increase on the refraction index of porous thin films as a result of the adsorption of airborne contaminants. Pareek et al. [15] found that oil vapour contamination is also responsible for the increase of the refractive index of porous AR coatings.

5.9   NANOMATERIAL-BASED SOLAR COOL COATINGS

The global building and construction industry is responsible for both 40% of the entire world's energy consumption and emitting a third of the world's greenhouse gases annually. At least 50% of the total energy consumed by this industry are for powering heating, ventilating and air-conditioning (HVAC) systems. Passive cooling and solar heat insulation technologies are often being regarded as solutions in addressing the global energy crisis, in which they are being considered as reducing or even consuming zero energy.

Solar radiation plays the main role in terms of buildings gaining heat when they are transmitted via the envelope. The heat will be trapped and increased inside the building. Buildings therefore are more likely to use nanomaterial-based solar cool coatings (NSCCs) in order to address the issue of excessive solar heat and energy consumption. These coatings are currently the most reliable in terms of passive cooling technologies. NSCCs are composite materials where it is made of thin-layered substrates mixed with nanosized additives, which is the primary component due to its distribution of solar reduction function onto a normal coating material. Binders are made of thin-layered substrates and they are added with nanosized additives to provide a coating to the surfaces of buildings where required.

NSCCs are widely used for the past few years as a solution to the high energy consumption in buildings. In Ref. [16], the authors have conducted a market research to show that the solar coatings will have a 70% in saving energy on a global scale from 2013 to 2019. The number of patents being filed is evidence to the increased attention and research being taken place to increase the use of NSCCs as well as proving that it is a pioneering technology in passive cooling. In Ref. [17], the authors have revealed that there is a significant increase by 38% between 2013 and 2015 on the number of patents being filed on smart window coatings. Meanwhile, the patents for thermal barrier coatings have increased by 32% between 2011 and 2015.

5.9.1 METAL-BASED NANOADDITIVES

For several decades, the key component in transparent coatings used in solar heat reflection is metal. Aluminium (Al), gold (Au), silver (Ag), copper (Cu) and platinum (Pt) have higher performance in terms of possessing high reflectivity and low absorptivity properties. Incorporating these metals into NSCCs leads to reflection of solar heat that would have otherwise penetrated into the indoor environment of the building. This passive cooling feature means that the indoor environment will need to use less energy consumption for cooling purposes. Significant amount of research was conducted on solar cool coatings. The least difficult in terms of implementation and usage are gold (Au) and silver (Ag).

Cher (2014) prepared nanogold (Au) films to be applied on the glass surface by using aerosol-assisted CVD method. Deposition was done inside a cold-walled horizontal-bed CVD reactor. The resulting nanogold layers possess various morphologies depending on the various reaction temperatures. Observation was made by placing a layer of gold nanoparticle at a temperature of 500°C as shown in Figure 5.6a. Figure 5.6b and c shows individual gold nanoparticles on the top plate of films being subjected to a temperature of 400°C. This shows that thermophoresis influences the increase of particle size, meaning that prior to deposition, the nanoparticles formation and aggregation of gold atoms occur in gas-phase reactions.

Chen et al. [18] conducted an alternative study where they fabricated ultra-smooth and ultrathin silver films using nanosilver (Ag) by utilising electron beam evaporation on glass substrates. The research has discovered that nanosilver films have high uniformity as well as major reduction of the size of nanoparticles at 3–8.5 nm. This is due to the thin germanium being used as the wetting materials as shown in Figure 5.6d. A nanosilver film without the thin germanium is shown in Figure 5.6e, and it clearly shows the irregular silver islands that are isolated on the substrate. Silver mirror reaction was simulated to synthesis the nanosilver. This means that the precursor silver nitrate will be chemically reduced using ammonia, glyoxal and triethanolamine; the last two act as reducing agents. The mean size of the resulting silver nanograins is 28 nm.

5.9.2 METAL OXIDE-BASED NANOADDITIVES

Metal oxide nanostructures have more stability while do not have oxidation issues when compared to other metals that are prone to oxidation, which have limited

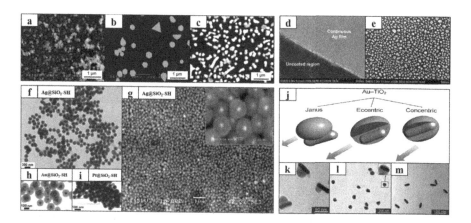

FIGURE 5.6 (a–c) SEM (scanning electron microscope) images of nano-Au deposited on the top plates at various temperatures. (d, e) SEM images of an Ag layer (d) with and (e) without a Ge wetting layer. (f–i) TEM (transmission electron microscope) and SEM images of Ag@SiO$_2$-SH (f, g), (h) Au@SiO$_2$-SH and (i) Pt@SiO$_2$-SH. (j–m) Schematic (j) of Au@TiO$_2$ nanorods with various geometries; TEM images of Au@TiO$_2$ nanorods with (k) Janus, (l) eccentric and (m) concentric geometries [17].

application. This has therefore garnered more research and investigation. More conventional applications and studies are conducted on metal oxide-based nanoadditives applied in NSCCs, which are zinc oxide (ZnO) and tin oxide (SnO$_2$). This is due to their remarkable heat insulative attributes in addition to chromogenic nanomaterials such as vanadium oxide (VO$_2$) and tungsten oxide (WO$_3$). They have shown potential in energy savings. Furthermore, an easier and efficient method is molecular doping, which could further enhance the solar reduction properties of these nanomaterials. These include using fluorine as an oxygen replacement and high valent metal.

NSCCs would possess solar reflection properties when incorporated with reflection-based metal oxide nanoadditives. The NSCCs application on building envelops such as glass or opaque walls has the capability of reflecting the solar heat via plasmon resonances between the incoming solar radiation and nanoadditives. There have been numerous papers being done on NSCCs that incorporates reflection-based metal oxide nanoadditives. The more renowned nanomaterials being researched are zinc oxide (ZnO) and tin oxide (SnO$_2$) and their doping composites.

Soumya [19] synthesised solar thermal control films by using nanosized zinc oxide (ZnO) into polymethylmethacrylate (PMMA) polymer matrix composite. Microwave-assisted polyol method was used for the synthesisation of ZnO nanoparticles. Subsequently, solution-casting technique was used to produce ZnO/PMMA nanocomposite films. The size range of ZnO nanoparticles is 50–100 nm with the nanomaterials having many types of morphologies including nanosheets, nano-needle, nanospheres, nanoplatelets and nanoflakes due to the various techniques and capping agents being used as shown in Figure 5.7a–f.

Higher valent elements including gallium (Ga), aluminium (Al) and indium (In) could be used as alternative to zinc (Zn) atoms for the purpose of achieving significant

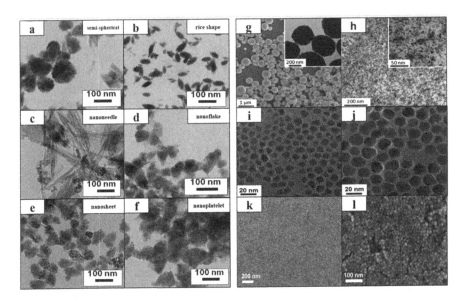

FIGURE 5.7 (a–f) TEM images of ZnO nanoparticles synthesised by (a–c) reflux method and (d–f) microwave method using different capping agents. (g, h) SEM and inserted TEM images (g) of the pristine GZO; TEM images (h) of GZO after disaggregation. (i–l) TEM images (i, j) of AZO nanocrystals obtained at various temperatures; SEM images (k, l) of AZO thin film under various magnifications [17].

changes in the ZnO nanoparticles optical properties. An example of this is gallium zinc oxide (GZO), where it has potential as a barrier of infrared radiation. Trenque [20] pioneered the synthesisation of micron-scaled particles through polyol-mediated precipitation method that encompasses forced hydrolysis of zinc acetate and gallium salt in diethylene glycol (DEG).

The finished product is nanosized GZO dispersion that is transparent, which was achieved by using the process of disaggregation within hexadecane-1-thiol (HDT) under ultrasonication. The as-obtained GZO nanocrystallites are dispersed uniformly at mean size of 11–12 nm, which is significantly smaller when compared to pure GZO with mean size up to 550 nm as shown in Figure 5.7g and h. Earth-abundant aluminium zinc oxide (AZO) is more cost-effective, less toxic and more eco-friendly when compared to other GZO nanomaterials. It is thus able to serve as a substitute for other transparent conducting oxide (TCO) materials and thus have garnered much attention in the past few years,

Buonsanti [21] developed a nanosynthetic method that could fabricate AZO nanocrystals where its size and doping levels can be controlled. Figure 5.7i and j shows two types of the as-obtained AZO. One is pseudo-spherical (10 nm in diameter obtained at 240°C) and hexagonal-shaped nanocrystals (15 nm in diameter obtained at 260°C). Figure 5.7k and l shows the uniform AZO films made possible by drop casting the AZO nanocrystals. Luo [22] conducted similar study in preparing the AZO/epoxy nanocomposites.

The synthesisation of AZO nanoparticles involved the calcination of a range of temperatures by using the homogeneous precipitation method. The nanoparticle size increases from 25 to 90 cm as the calcination temperature increases from 200°C to 700°C. This shows that AZO nanoparticles calcinate at comparatively higher temperatures before aggregating into clusters with the naturally relatively larger sizes. The completed product is the AZO/epoxy nanocomposites, which is simply an AZO dispersion with mean diameter of 65 nm, into curing agent and transparent epoxy resin before being heated at 130°C for 2 hours.

5.9.3 Absorption-Based Nanoadditives

An alternative to solar reflection is solar absorption mechanism in terms of passive cooling strategy in decreasing solar heat penetration into the building's glazed envelopes. When compared to the reflection-based nanoadditives, it is possible for absorption-based nanoadditives to be merged with NSCCs prior to applying them on the inner surface of the double-glazed units' outer pane. The two panes would have an air gap which forms a protection for the nanoadditives and NSCCs from degradation and oxidation due to its exposure to the ambient environment. In addition, it also acts as a heat barrier, preventing it from entering through the glass, thus in practical setting, inside a building.

Tungsten oxide (WO_3) and alkali-metal-doped WO_3 are the most renowned nanomaterials. They are commonly used to produce smart windows as they provide infrared radiation attenuation via infrared absorption. Furthermore, both tungsten oxide (WO_3) and alkali-metal-doped WO_3 possess unique electrochromic properties. This feature responds to voltage by changing the optical properties, which means it could inhibit the light and heat to pass through.

5.9.4 Metalloid-Based Nanoadditives

Nanomaterials that are highly transparent solar filters for NSCCs are not limited to metal and metal oxide as other nanoadditives could also be used. An example of this is lanthanum hexaboride (LaB6), which is similar to metal. It is capable of absorbing infrared rays via surface plasmon resonance. Jiang [23] conducted a study on preparing stabilised water-dispersed LaB6 nanoparticles using cetyltrimethylammonium bromide (CTAB). Subsequently, the nanocomposite films were synthesised through spin-coating a solution of sol-gel silica with LB6 nanoparticles on glass substrates.

LB6 nanoparticles that are untreated has the average size ranging from 700 to 1200 nm. In contrast, LaB6 nanoparticles that are stabilised via CTAB has smaller size averaging from 70 to 250 nm. To raise its performance further, other solar absorptive nanomaterials were added along with LB6 nanoparticles during application. Tang accomplished this by using a solution-casting method to prepare the LB6-ITO nanocomposite films by doping and mixing LB6 (80–120 nm in diameter) and ITO (20–30 nm in diameter) nanoparticles into polyvinylbutyral (PVB) matrix. The PVB matrix were added with both LB6 and ITO nanoparticles due to their remarkable adhesion property and strong interfacial bonding to the PVB molecules.

5.10 KEY OF BUILDING APPLICATIONS

NSCCs coatings and films are widely used in transparent windows and facade glazing. This is because they are subjected to solar radiation; thus, such coating could be used as a form of control and for the purpose of passive cooling. NSCCs optical properties pertain to IR reflectance/absorptance and visible light transmittance, both of which are essential for thermal environment and indoor lighting. The definition of indoor temperature reduction is the decrease of the temperature when NSCCs is being applied. It is a success when the optical property is performing accordingly in actual settings, particularly the infrared reflectance/absorptance.

5.10.1 GOLD (AU)

This section discusses metals that are most suitable for the purpose of passive cooling effect: gold and silver could reflect solar hear irradiation. Chew [24] studied the nanogold (Au) heat reflective application on solar cool films. The films were made using the aerosol-assisted CVD method. Due to the nanogold metallic reflectance's properties, the final product is nanocomposite films that are transparent and have a light blue tint and acceptable amount of optical transparently. The most essential element is its wide reflectance ranging from 500 nm to long wavelength of a few micron region as shown in Figure 5.8a.

5.10.2 ZINC OXIDE (ZNO) AND ALUMINIUM ZINC OXIDE (AZO)

Metal oxides have greater stability when compared to other metal-based nanomaterials used for NSCCs when subjected to direct exposure to the ambient environment. Due to these advantages, NSCCs applied on glazed building envelopes commonly use metal-oxide-based nanomaterials due to their direct exposure to ambient environment particularly in climates that are extremely hot. An example of this is when Soumya [25] showed nanocomposite films made with nano zinc oxide (ZnO) with the NIR reflectance property. Solution-casting technique was used to produce nanocomposite films made with ZnO/PMMA/PU.

NIR reflectivity of the nanocomposite films incorporated with ZnO nanocrystalline enhanced to 1100 nm or 55% as shown in Figure 5.8b. Its possibilities are still vast in terms of its role as solar thermal control interface as solar films are used in buildings despite its partial loss of visible lights. Adding aluminium (Al) into ZnO results in AZO, where it is more cost-effective for heat reduction solutions in buildings when compared to ITO and ATO. Buonsanti [21] has made a successful attempt at using AZO nanocrystals to produce uniform films, by dropping casting AZO nanocrystals from a dispersion in 85:15 hexane/octane.

These as-prepared films with the transmittance lower than 60% at 1800 nm in the NIR region are very useful as the lower cost and an environmental friendly alternative to ITO nanocoatings in the application of smart windows. It is a suitable replacement to other types of ITO nanocoatings used in smart windows. Soumya [25] fabricated AZO-embedded PMMA nanocoatings using the same method in fabricating ZnO/PMMA nanocomposite films. Dip coating involves dipping of glass into AZO/PMMA colloid to produce nanocoating. Similarly, layer-by-layer (LbL)

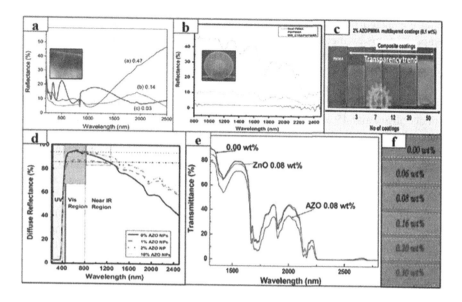

FIGURE 5.8 (a) Optical property of nanoparticle Au-SnO$_2$ composite films with inserted sample photograph. (b) Optical property of ZnO/PU/PMMA sheets fabricated via various methods with inserted photograph of ZnO/PU/PMMA sheet. (c) Optical property photograph. (d) AZO/PMMA nanocoatings photograph. (e, f) Optical property (e) and photograph (f) of AZO nanocomposites [17].

coating technique, as the name implies, is used to fabricate multilayers as shown in Figure 5.8c. These as-prepared nanocoatings have both NIR reflectivity and ultraviolet (UV) shielding properties as shown in Figure 5.8d.

Temperature can be reduced by 7°C by glass that is coated with transparent clear AZO/PMMA solar. This result is highly potential in addressing energy consumption issues by buildings. Luo [22] studied the mixing of as-prepared AZO nanoparticles with the transparent diglycidyl ether of bisphenol A (DGEBA) epoxy resin. Introduction of AZO nanoparticles at just 0.08 wt% results in approximately 10% improvement upon the infrared light shielding efficiency within the NIR region of 1300 nm as shown in Figure 5.8e.

Figure 5.8f shows the as-prepared transparent AZO nanocomposites with competing potentials in terms of incorporating them into infrared shielding windows and heat mirrors. Similarly, Ni et al. [26] employed the radio frequency (RF) magnetron sputtering technique that used AZO nanoparticles to fabricate films that are transparent and have high infrared reflection. The infrared cut-off rate and the highest reflection are at 91% at 1700 nm wavelength.

5.10.3 Indium Tin Oxide (ITO) and Antimony Tin Oxide (ATO)

When tin oxide is doped with either indium (In) or antimony (Sb), the result is ITO and ATO, respectively. These two elements are considered essentials in their application onto materials with solar blocking properties in buildings with many researches being conducted thus far. Liu [27] managed to produce highly transparent nano-composite films by means of a two-step method. The resulting ITO has a diameter of 6.3 nm and is produced by coasting the glass substrates with the mixture solution before drying it at 60°C temperature for 4 hours.

Figure 5.9a shows the finished product whereupon the NIR radiation cuts off at 50% with a length of 1500 nm. At least 85% of visible light could pass through the film at 550 nm and 45% UV ray is being blocked at 350 nm. It is suitable for optical applications particularly in buildings with transparent functional coatings. Jiang [28] fabricated nanocomposite films that are transparent and infrared blocking proper-ties by using ultrasonic-assisted LbL. The films that Jiang [28] used reduce infrared radiation by 80% at 1600 nm above as shown in Figure 5.9b. This shows that it is suitable to be applied in buildings' green windows.

Tao [29] used ITO and epoxy to fabricate visibly transparent nanocomposites. First, DGEBA epoxy resin and agent trimethyl-1,6-hexanediamine (curing agent) is mixed with grafted ITO. The glass was then coated and cured before leaving it overnight at the temperature of 80°C. The completed films consist of ITO/epoxy nanocomposite that has near total NIR shielding efficiency at 60 wt% with the ITO wavelength to be at least 1500 nm as shown in Figure 5.9c. It is used as a transparent and UV/IR opaque optical coatings. A study conducted by Llordés [30] meanwhile incorporates ITO nanocrystals into niobium oxide (NbOx) glass through dynamic control of solar radiation transmittance.

Figure 5.9d shows that the as-obtained films that are transparent has the capability to selectively block NIR light while its maximum NIR shielding is at 50% as a result of its electrochromic property. In terms of ATO nanoadditives, Li [31] incorporated the glass with ATO nanocrystal, in doing so achieving a thin film that is transparent. Waterborne polyurethane (WPU) is a stabilised suspension and it is incorporated with as-prepared ultrasonically pre-treated ATO nanocrystals. Next, the suspension is being casted uniformly on glass slide substrate before leaving it to dry at a tem-perature of 60°C for an hour. The finished product improved the heat reduction by 48.3% when compared to fluorine tin oxide (FTO) and ITO nanofilms as shown in Figure 5.9e.

The NIR blocking properties of these films unlock its possibility to be used in smart windows. Further assessment was carried out on the ATO nanofilm's thermal performance. Figure 5.9f and g show that these films are more efficient in terms of providing solar shielding, which is up to the temperature of 4.5°C when compared to ITO and FTO glasses. Gao [32] studied the impact on applying vanadium oxide (VO_2) onto ATO nanoparticles for the fabrication of composite thermochromic smart glass foil.

The process of producing nanocomposite foil is to first prepare, mix and stir uni-formly of VO_2, ATO and PU before casting resultant suspension on polyethylene-terephthalate (PET) substrates and leaving it to dry at a temperature of 90°C for 60 seconds. The resulting as-prepared foils possess solar shielding and remarkable solar modulation properties. Figure 5.9h shows a decrease in solar transmittance by 20% when ATO content is at 9%, which is considered a breakthrough for smart windows. The foil was then applied to a house model in order to evaluate its infrared shielding property. The result showed that it is possible to improve 17°C (compared against

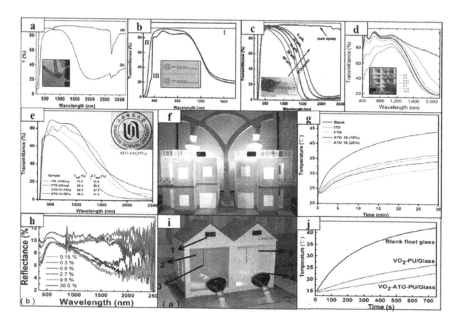

FIGURE 5.9 (a) Optical property of ITO nanocomposite film with inserted sample photograph. (b) Optical property of quartz substrates with ITO composite films with inserted sample photograph. (c) Optical property of ITO/epoxy coatings with various concentrations of ITO with inserted photograph of samples. (d) Optical property of ITO-in-NbOx film on a glass substrate with inserted sample photograph. (e–g) Optical property (e) of ATO nanofilm with inserted sample photograph; photograph (f) of building models as testing system; temperature curves (g) of various nanofilms. (h–j) Optical properties (h) of VO_2-ATO-PU nanocomposite foil with various ATO contents; photograph (i) of testing system; temperature curves (j) of various coated glasses [17].

blank float glass) and 3.5°C (compared against VO_2-PU glass) temperature depression with 30% addition of ATO as shown in Figure 5.9i and j.

5.10.4 VANADIUM OXIDE (VO_2)

Vanadium oxide (VO_2) is one of the renowned solar reflective nanoadditives that is commonly used for the buildings' passive cooling purposes due to its thermochromic materials. Gao et al. [33] have fabricated VO_2-based nanocomposite films, which are reported to be stable, transparent and flexible. The VO_2 nanoparticles were coated with silica before being mixed using ultrasonic method along with PU and silane couple. The mixture is stirred for an hour before making a cast on PET substrates and drying to obtain the finished product.

The completed film has remarkable temperature-responsive thermochromic in NIR region. It has 13.6% solar modulation efficiency thus making it suitable for new applications in NIR modulation of glass that are used within the construction industry. Li [31] used titanium oxide (TiO_2) as the coating when synthesising multifunctional coatings containing $VO_2@TiO_2$ core–shell structures. The coating was synthesised through casting the suspension comprising of $VO_2@TiO_2$ nanoparticles onto a float glass substrate. The colour of the finished product is yellowish-brown and reached a maximum of 75.2% NIR shielding at 2000 nm. This means that it has potential application in energy-efficient smart coatings.

5.10.5 TUNGSTEN OXIDE (WO_3) AND ALKALI METAL-DOPED TUNGSTEN OXIDE ($AxWO_3$)

Tungsten oxide (WO_3) and alkali metal-doped WO_x are able to switch their states depending on the dynamic of the voltage. This is unique when compared to VO_2 where it only reacts to a change in temperature. WO_3 and WO_x therefore have vast potential as a material for electrochromic windows. Guo et al. [34] have produced a rod with NIR shielding property using W18O49 nanoparticles. The collodion–ethanol mixed solution is mixed with as-prepared W18O49 nanopowders before being applied on quartz glass. Its NIR shielding was subsequently assessed.

Observations have been made on W18O49 coated glass' ability to significantly absorb the NIR light, but at the same time, its visible transparency remains high. Such glass is ideal for optical-related glass as shown in Figure 5.10a and b. Simulation was conducted in order to evaluate the solar heat filters of the W18O49 nanorods. The results returned that glass with W18O49 coating demonstrate the ability to maintain inner temperature at 25.8°C after being subjected to an hour of irradiation, which is significantly lower when compared to ITO glass as shown in Figure 5.10c–e. Guo et al. [35] conducted a similar study on a quartz glass coated with caesium tungsten oxide ($CsxWO_3$). The glass displays a remarkable infrared absorption property in a waveband ranging from 800 to 1500 nm as shown in Figure 5.10f. The synthetisation of $CsxWO_3$ films involved the painting coating slurry to be directly applied on quartz glass via an applicator.

$CsxWO_3$-based glass further decrease the temperature when compared to quartz glass (by 10°C) and ITO glass (by 6°C) through its heat ray absorption property as shown in Figure 5.10g. $CsxWO_3$ is ideal to be used for smart windows. Kang et al. [36] have made a similar study of fabricating solar control foil by using ammonium (NH_3) instead of caesium (Cs). Ultrasonic dispersion of ammonium tungstate [$(NH_4)xWO_3$] powders was done within deionised water with polyvinylpyrrolidone (PVP). The formation of the film occurred and was casted on the PET substrate uniformly.

The $(NH_4)xWO_3$ nanocrystallites-based foils could maintain a 67.6% transmission of visible light while able to cut off NIR radiation by 65.8%. Such materials will prove to be a useful filter that is energy efficient. Wu (2015) synthesised smart multifunctional coatings through the use of $CsxWO_3/ZnO$ nanocomposite. These coatings have heat insulation and photocatalytic clean-up properties.

The preparation of nanocomposite coatings necessitates the dispersion of $CsxWO_3$ and ZnO nanopowders into a mixed solution containing collodion–ethanol. An applicator was used to drop them onto quartz glass. The results have shown that $CsxWO_3$ have remarkable NIR light shielding with a range of 780–2600 nm, which is 22% higher when compared with ITO nanocoatings within the wavelength of 780–1450 nm. Under the same comparison, the temperature depression of $CsxWO_3$ is

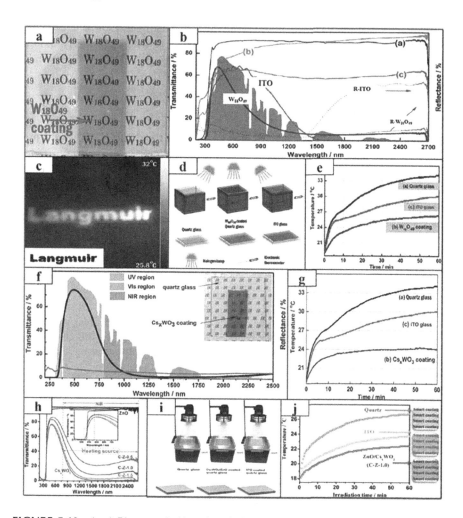

FIGURE 5.10 (a–e) Photograph (a) and optical properties (b) of W18O49 film coated on quartz glass; thermographic image (c) of W18O49 nanopowders; schematic (d) of simulated experiment; temperature profiles (e) dependent on irradiation time. (f, g) Optical properties (f) of CsxWO3 films coated on quartz glass (black-transmittance, red-reflectance) with inserted sample photograph; temperature profiles (g) dependent on irradiation time. (h–j) Optical properties (h) of CsxWO3, ZnO and hybrid CsxWO3-ZnO films; schematic (i) of simulated heat ray shielding test; temperature variation profiles (j) with light irradiation time as well as photographs of various sample glasses [17].

at 2°C in heat ray shielding test as shown in Figure 5.10h–j. These results mean that CsxWO3/ZnO smart nanocoatings have vast potential in terms of energy savings use in buildings.

In an alternate study to CsxWO3 and ZnO combination, Zeng synthesised high-performance NIR shielding nanocomposite coatings by using a combination of SiO_2 and CsxWO3 nanoparticles. The preparation involved mixing CsxWO3 and SiO_2

with tetraethoxysilane (TEOS) before using the spin-coating technique to drop them to the glass. The as-obtained nanocoatings possess NIR shielding property and have a maximum 70% improvement when compared to ITO glass with excellent optical stability. It has vast potential within the architecture field.

5.11 SUMMARY

In response to the outlined objectives, it is foreseeable that NSCCs will play a more important role as a promising passive cooling technology in the future to reduce solar heat gain into buildings and lower building energy consumption. In the context of global climate change and fast urbanisation and due to energy deficiency and environment deterioration, the development of NSCCs will be fast, mainly including two aspects, large-scale synthesis of high-performance nanomaterials and cost- and time-efficient coating fabrication techniques.

REFERENCES

1. Pariafsai, F. A review of design considerations in glass buildings. *Frontiers of Architectural Research*, 2016, **5**(2): pp. 171–193.
2. Manz, H., et al. Series of experiments for empirical validation of solar gain modeling in building energy simulation codes—Experimental setup, test cell characterization, specifications and uncertainty analysis. *Building and Environment*, 2006, **41**(12): pp. 1784–1797.
3. Strachan, P. Model validation using the PASSYS test cells. *Building and Environment*, 1993, **28**(2): pp. 153–165.
4. Bui, V., et al. Evaluation of building glass performance metrics for the tropical climate. *Energy and Buildings*, 2017, **157**: pp. 195–203.
5. Crawley, D.B., et al. EnergyPlus: Creating a new-generation building energy simulation program. *Energy and Buildings*, 2001, **33**(4): pp. 319–331.
6. Schodek, D.L., P. Ferreira, and M.F. Ashby. *Nanomaterials, Nanotechnologies and Design: An Introduction for Engineers and Architects*, 2009. Butterworth-Heinemann, USA.
7. Bloomfield, D. An overview of validation methods for energy and environmental software. *ASHRAE Transactions*, 1999, **105**: p. 685.
8. Abdin, A.R., A.R. El Bakery, and M.A. Mohamed. The role of nanotechnology in improving the efficiency of energy use with a special reference to glass treated with nanotechnology in office buildings. *Ain Shams Engineering Journal*, 2018, **9**(4): pp. 2671–2682.
9. Glöß, D., et al. Multifunctional high-reflective and antireflective layer systems with easy-to-clean properties. *Thin Solid Films*, 2008, **516**(14): pp. 4487–4489.
10. Liu, Z., et al. Sol–gel SiO_2/TiO_2 bilayer films with self-cleaning and antireflection properties. *Solar Energy Materials and Solar Cells*, 2008, **92**(11): pp. 1434–1438.
11. Prado, R., et al. Development of multifunctional sol–gel coatings: Anti-reflection coatings with enhanced self-cleaning capacity. *Solar Energy Materials and Solar Cells*, 2010, **94**(6): pp. 1081–1088.
12. Helsch, G. and J. Deubener. Compatibility of antireflective coatings on glass for solar applications with photocatalytic properties. *Solar Energy*, 2012, **86**(3): pp. 831–836.
13. Helsch, G., et al. Adherent antireflection coatings on borosilicate glass for solar collectors. *Glass Technology – European Journal of Glass Science and Technology Part A*, 2006, **47**(5): pp. 153–156.

14. Cathro, K., D. Constable, and T. Solaga. Durability of porous silica antireflection coatings for solar collector cover plates. *Solar Energy*, 1981, **27**(6): pp. 491–496.

15. Pareek, R., et al. Effect of oil vapor contamination on the performance of porous silica sol-gel antireflection-coated optics in vacuum spatial filters of high-power neodymium glass laser. *Optical Engineering*, 2008, **47**(2): p. 023801.

16. Surekha, K. and S. Sundararajan. Self-cleaning glass. In: *Anti-Abrasive Nanocoatings*, 2015. Elsevier, UK: pp. 81–103.

17. Zheng, L., T. Xiong, and K.W. Shah. Transparent nanomaterial-based solar cool coatings: Synthesis, morphologies and applications. *Solar Energy*, 2019, **193**: pp. 837–858.

18. Chen, W., et al. Ultra-thin ultra-smooth and low-loss silver films on a germanium wetting layer. *Optics Express*, 2010, **18**(5): pp. 5124–5134.

19. Soumya, S., et al. Near IR reflectance characteristics of PMMA/ZnO nanocomposites for solar thermal control interface films. *Solar Energy Materials and Solar Cells*, 2014, **125**: pp. 102–112.

20. Trenque, I., et al. Visible-transparent and UV/IR-opaque colloidal dispersions of Ga-doped zinc oxide nanoparticles. *New Journal of Chemistry*, 2016, **40**(8): pp. 7204–7209.

21. Buonsanti, R., et al. Tunable infrared absorption and visible transparency of colloidal aluminum-doped zinc oxide nanocrystals. *Nano Letters*, 2011, **11**(11): pp. 4706–4710.

22. Luo, Y.-S., et al. Preparation and optical properties of novel transparent Al-doped-ZnO/epoxy nanocomposites. *The Journal of Physical Chemistry C*, 2009, **113**(21): pp. 9406–9411.

23. Jiang, F., et al. Uniform dispersion of lanthanum hexaboride nanoparticles in a silica thin film: Synthesis and optical properties. *ACS Applied Materials & Interfaces*, 2012, **4**(11): pp. 5833–5838.

24. Chew, C., et al. Aerosol-assisted deposition of gold nanoparticle-tin dioxide composite films. *RSC Advances*, 2014, **4**(25): pp. 13182–13190.

25. Soumya, S., et al. Enhanced near-infrared reflectance and functional characteristics of Al-doped ZnO nano-pigments embedded PMMA coatings. *Solar Energy Materials and Solar Cells*, 2015, **143**: pp. 335–346.

26. Ni, J., Q. Zhao, and X. Zhao. Transparent and high infrared reflection film having sandwich structure of SiO_2/Al: ZnO/SiO_2. *Progress in Organic Coatings*, 2009, **64**(2–3): pp. 317–321.

27. Liu, H., et al. A simple two-step method to fabricate highly transparent ITO/polymer nanocomposite films. *Applied Surface Science*, 2012, **258**(22): pp. 8564–8569.

28. Jiang, C., et al. Fabricating transparent multilayers with UV and near-IR double-blocking properties through layer-by-layer assembly. *Industrial & Engineering Chemistry Research*, 2013, **52**(37): pp. 13393–13400.

29. Tao, P., et al. Preparation and optical properties of indium tin oxide/epoxy nanocomposites with polyglycidyl methacrylate grafted nanoparticles. *ACS Applied Materials & Interfaces*, 2011, **3**(9): pp. 3638–3645.

30. Llordés, A., et al. Tunable near-infrared and visible-light transmittance in nanocrystal-in-glass composites. *Nature*, 2013, **500**(7462): pp. 323–326.

31. Li, Y., et al. Tunable solar-heat shielding property of transparent films based on mesoporous Sb-doped SnO_2 microspheres. *ACS Applied Materials & Interfaces*, 2015, **7**(12): pp. 6574–6583.

32. Gao, Y., et al. Nanoceramic VO_2 thermochromic smart glass: A review on progress in solution processing. *Nano Energy*, 2012, **1**(2): pp. 221–246.

33. Gao, Y., et al. Enhanced chemical stability of VO_2 nanoparticles by the formation of SiO_2/VO_2 core/shell structures and the application to transparent and flexible VO_2-based composite foils with excellent thermochromic properties for solar heat control. *Energy & Environmental Science*, 2012, **5**(3): pp. 6104–6110.

34. Guo, C., et al. Discovery of an excellent IR absorbent with a broad working waveband: CsxWO₃ nanorods. *Chemical Communications*, 2011, **47**(31): pp. 8853–8855.

35. Guo, C., et al. Facile synthesis of homogeneous CsxWO₃ nanorods with excellent low-emissivity and NIR shielding property by a water controlled-release process. *Journal of Materials Chemistry*, 2011, **21**(13): pp. 5099–5105.

36. Kang, L., et al. Transparent (NH₄) xWO₃ colloidal dispersion and solar control foils: Low temperature synthesis, oxygen deficiency regulation and NIR shielding ability. *Solar Energy Materials and Solar Cells*, 2014, **128**: pp. 184–189.

6 Air Nano Purification

6.1 INTRODUCTION

The health of a building's occupants can be substantially influenced by the indoor air quality which is of particular relevance to healthy buildings that strive to provide a healthy environment for their inhabitants [1]. The subject matter of indoor quality has received increasing worry and attention as people spend a majority of their time indoors. Outdoor, physical systems, toxicological, indoor and microbiological ventilations are among the elements impacting indoor air quality [2]. The significant pollution stemming from volatile organic compounds (VOCs) is among the side effects that the implementation of synthetic building materials has exerted upon the indoor environment, partially as a result of the developments in construction technology. Temperatures between 50°C and 260°C represent the boiling point of VOCs [3]. The toxicity of the VOCs is evident to both humans and the environment, where such substances generate fine particulates in the atmosphere and can add to the ozone formation [4]. Sick building syndrome, impaired neurobehavioural function, respiratory diseases, and other such chronic and severe health effects can be directly caused by VOC exposure [2].

Oxidation methods, adsorption methods or an amalgamation of the two comprise the principal mechanisms put forward to eliminate VOCs [5,6]. VOCs are transferred from the air to the solid phase via fibre, biochar, activated carbons or other such adsorbents in the adsorption technique [7–9]. Pore blockages and saturation are among the difficulties that this technique encounters. In comparison, increased degradation activity towards polar VOCs (OVOCs > Ahs > AlHs) is demonstrated by the oxidation method which represents a more cost-effective manner of removing VOCs [10]. Among the most frequent oxidation methods are thermal and photocatalytic oxidation. The efficacy of oxidation increases in tandem with temperature in the case of thermal oxidation reactions that necessitate temperature superseding 600°C. As shown in Figure 6.1, organic compounds in indoor air are transformed into odourless and harmless carbon dioxide (CO_2) and water vapour for air purification via the use of ultraviolet (UV) light and non-semiconductor catalysts within the most frequent method of photocatalytic oxidation [5,11,12].

The indoor air in novel renovated buildings is full of halocarbons, aldehydes and aromatics which are VOCs [14]. Coverings were determined to be the principal source of VOCs and exposure levels to VOCs were found to be equivalent in multiple buildings and indoor materials, as signified by the measurements [15]. Construction projects prevalently utilise VOCs, and VOCs are widespread in the industry [5]. In the case of freshly decorated or constructed buildings, the standard ambient levels of VOC could be well surpassed as many VOCs convert a substantial proportion of their volume into gas in a short period of time [2]. Using a radiant floor-heating system to fumigate a housing unit with heat is one possible manner of diminishing

DOI: 10.1201/9781003281504-6

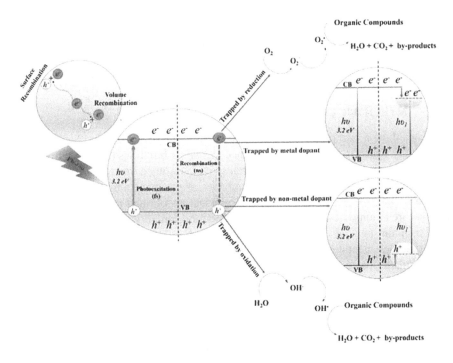

FIGURE 6.1   Mechanisms of photocatalytic oxidation for the removal of VOCs [13].

indoor concentrations of VOC alongside its emissions [16]. Emissions and solvents of VOC could be substantially diminished via the heat of the heating system being implemented for a sufficiently long time [17]. The efficacy, sanitation and aesthetics of the buildings will be positively impacted by applying paint, a coating or mortar synthesised with these using TiO_2, thereby implementing photocatalysis [18–20]. Thus, construction materials should be an area attentively focused upon.

The materials for the sensing and removal of VOCs have been extensively researched and summated in the literature over the last few decades [5,21–23]. Silica-nanosphere-based materials, graphene-based materials, TiO_2, zinc indium sulphide and other such materials formed the focal point of the review of the literature centring on such particular categories of materials [24–29]. Similarly, a summation of the outlooks of multiple VOCs, visible light and low temperatures was among the particular contexts undergoing catalytic oxidation processes that were focused upon [5,30–32]. Nevertheless, taking into account applications on buildings in tandem with a material perspective has sparsely been researched. Given that the extant literature includes extensive reviews of adsorption-based materials, this study does not analyse such materials [5,9]. Physiochemical features served as the basis to stratify multiple materials given that the performance of catalytic oxidations varies between materials which, in turn, impacts their potential implementation on buildings. Hence, as seen in Figure 6.2, the application on buildings of the materials for

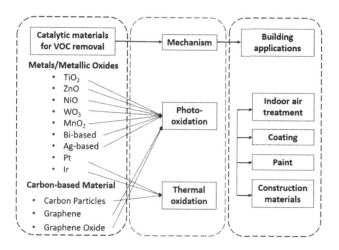

FIGURE 6.2 Outline of this chapter.

catalytic removal of VOCs is summarised in this chapter. Comprehensive explications will be given for the applications of performance, synthesis and morphology. The literature on the implementation of photocatalysis materials on buildings will be expanded via this chapter.

6.2 METALLIC OXIDES

6.2.1 TITANIUM DIOXIDE (TiO_2)

It is widely acknowledged that the most applicable and efficacious material is TiO_2 due to its low cost, robustness against photocorrosion, wide band-gap energy and low toxicity [33]. The discovery of photoelectrochemical decomposition of water subjected to irradiation with light on TiO_2 occurred in 1972 [34]. Over many decades, the photocatalysis performances of TiO_2 and its offshoots have been investigated.

The treatments, morphology and structure of the TiO_2 particles enable the evaluation of photocatalyst performance [35,36]. At the primary particle size of 7 nm, a maximum is displayed in the trichloroethylene degradation over TiO_2, as per the analysis of how the morphology and size of TiO_2 particles are impacted by distinct synthesis parameters conducted by Maira et al. [35]. The different manners in attaining nanosized TiO_2 particles displayed by the distinct treatments implanted on amorphous TiO_2 precursors were compared in the mentioned study [36]. An increased adsorption ability on benzaldehyde enhanced photoactivity, and a greater number of hydrogen-bonded hydroxyl groups with better stability under RT outgassing were demonstrated by the TiO_2 treated with the hydrothermal method, when compared to the anatase TiO_2 treated with the thermal technique.

Distinct photocatalytic performances were displayed by TiO_2 with distinct morphologies. An enhanced efficacy for toluene adsorption was demonstrated by a new nanostructured gas filtering system containing TiO_2 thin films utilising atomic layer deposition (ALD) for VOCs in a certain academic work [37]. During the recurrent cycles of photocatalytic degradation of acetaldehyde and gaseous toluene, the photocatalytic activities of TiO_2 nanoparticles (TNP) film and TiO_2 nanotubes (TNT) were compared in a separate study [38]. The TEM images of fresh TNT and TNP are illustrated in Figure 6.3a and b. As demonstrated in Figure 6.3c and d, due to the repetition of the photocatalysis in a more oxidising atmosphere, TNP experienced rapid deactivation while five cycles of toluene degradation led to merely a moderate decrease in TNT photocatalytic activity. The aggregation of carbonaceous residues is averted in the TNT surface via the ease with which the TNT can provide O_2 molecules to the active sites with reduced mass transfer limitation and its highly structured open channel. During the photocatalytic degeneration of aromatic compounds, catalyst deactivation is averted by the highly beneficial structural features of TNT. When compared to TNT, increased durability and activity were observed regarding photocatalytic degradation of toluene and gaseous acetaldehyde in synthesised freestanding doubly open-ended TiO_2 nanotubes (DNT) film, per another academic work [39]. 1.3 and 1.8 times more activity for VOC degradation was observed in bare TNT and bare DNT, respectively, in the scenario where TiO_2 nanoparticles are added to the inner wall of the freestanding DNT film (NPDNT). Nevertheless, when compared to bare TNT, reduced activity was witnessed with TiO_2 nanotubes loaded with TiO_2 nanoparticles.

Rutile, brookite and anatase comprise the three crystal phases in which TiO_2 exists. The greatest practicality for environmental applications and photoactivity is seemingly possessed by the anatase form [25]. Significant improvements to photocatalytic performance were observed upon the effective synthesis of nanostructured brookite following a period where brookite was not considered to be an applicable

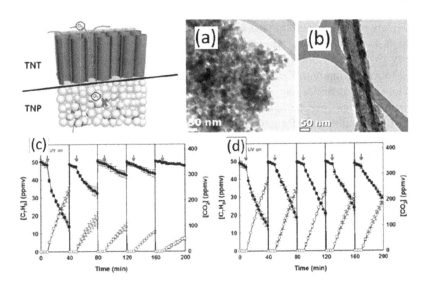

FIGURE 6.3 TEM images of fresh. (a) TNP. (b) TNT; repeated photocatalytic degradation cycles of gaseous toluene on. (c) TNP. (d) TNT in the air ([Toluene], O [CO_2]) [38].

photocatalyst [25,40,41]. Within gas-phase photocatalytic oxidations of methanol and hexane, the synergetic effect between rutile nanoparticles and anatase was examined in the work by Wu et al. [42]. If close contact was experienced by the rutile and anatase particles, this synergetic effect could be more impactful. The single-phase anatase TiO_2 benefits from the stable nature of the photocatalyst activity, as demonstrated in the long-term experiment that cannot be enhanced via sulphation. Bicrystalline TiO_2 exhibits less durability and photocatalytic activity towards gaseous toluene than tricrystalline TiO_2 [43]. A low-temperature hydrothermal route with HNO_3 was used to synthesise rutile, brookite and anatase tricrystalline TiO_2 in order to remove toluene cost-effectively and resourcefully from indoor air, per the same study [43]. A prevalently utilised benchmark model of photocatalysts with coexisting rutile and anatase phases is reflected by the $RHNO_3$ value of 0.8 (3.7% rutile, 15.6% brookite, 80.7% anatase) which, upon comparison to P25 TiO_2, demonstrated 3.85-fold higher photocatalytic activity and was the highest figure reached, per Figure 6.4a. Furthermore, Figure 6.4b illustrates that following five reuse cycles, substantial degradation was not witnessed despite the high levels of activity.

Nonetheless, the inherent wide band gap in conditions of visible light activation led to low photocatalytic activity in TiO_2, despite it being the dominant semiconductor photocatalyst. Thus, carbonaceous nanomaterials, nanomaterials mixed with distinct metal oxides and other such materials are used to alter TiO_2 to make it effective in the visible light region [44–47]. In the context of toluene degradation, urea–glass synthesis was enacted to generate a photocatalytic oxidation material responsive to visible light, forming the TiNbON compound with a band energy of 2.3 eV, per a separate academic work [44]. Adequate levels of stability and durability were observed in the visible light-driven catalysts that produced less formaldehyde than commercial TiO_2 and eliminated up to 58% of the toluene in the air, per the experimental findings of the mentioned paper. As illustrated in Figure 6.5a, a stellar risk-mitigating material for indoor environments was produced following the grafting of nano-CuxO clusters on TiO_2, as per the work of Qiu et al. [45]. The CuI species enable the presence of

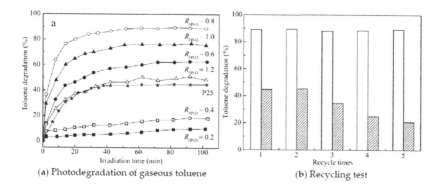

(a) Photodegradation of gaseous toluene   (b) Recycling test

FIGURE 6.4 Comparison results between TiO_2 and P25 for (a) photodegradation rate of gaseous toluene. (b) Recycling test over tricrystalline TiO_2 0.8 (blank) and P25 (filled) for five repeat uses [43].

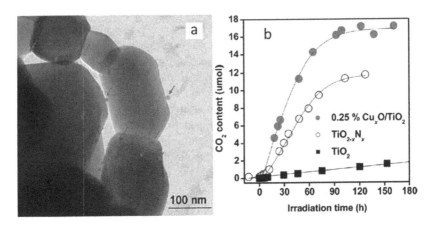

FIGURE 6.5 (a) TEM images of the 0.25% CuxO/TiO$_2$ sample. CuxO clusters (marked by red arrows) were highly dispersed on the TiO$_2$ surfaces. (b) Comparative studies of CO$_2$ generation over bare TiO$_2$, TiO$_2$-xNx and 0.25% CuxO/TiO$_2$ samples under the same conditions [45].

antimicrobial characteristics in dark conditions which, in turn, allow the TiO$_2$ to protoxidise VOCs in visible light effectively through its CuxO and CuII clusters. Hence, with suitable proportions of CuII and CuI and CuxO, a hybrid of TiO$_2$ and CuxO nanocomposites could enable the efficacious decrease of antipathogenic activity and VOCs. In conditions of long-term light irradiation, lower levels of stability and quantum efficiencies are found in the TiO$_2$-xNx sample than the higher levels demonstrated by the CuxO and TiO$_2$ hybrid sample, as illustrated in Figure 6.5b.

To remove VOCs in indoor environments, a distinct study adopted an annular reactor coated with nitrogen (N)-doped TiO$_2$ and unaltered TiO$_2$ [48]. In conditions of standard indoor comfort ranging between 50% and 60% within less humidified environments, over 90% of the target compounds o.m.p-xylenes and ethylbenzene were removed, demonstrating higher efficacy in the photocatalytic method utilising N-doped TiO$_2$ than the unaltered version. The capacity for toluene removal was assessed for modified TiO$_2$ nanoparticles with fluoride and Pt in a separate research paper [46]. During repeated cycles, deactivation of the Pt/TiO$_2$ could occur rapidly despite demonstrating increased photocatalytic degradation activity than unmodified TiO$_2$. As portrayed in Figure 6.6, the highest level of durability and photocatalytic activity for toluene degradation was demonstrated by F-TiO$_2$/Pt. Figure 6.7a illustrates the incorporation of Pt-rGO-TiO$_2$, a hybrid nanomaterial presented in another academic work [47]. Under IR irradiation, the VOCs are effectively decomposed due to the highly active photo-thermal responsive catalyst possessing a wide light wavelength absorption of 800–2500 nm. During the yield of CO$_2$ and the conversion of toluene, the efficacy of Pt-rGO-TiO$_2$ composites can be impacted by the light intensity, as displayed in Figure 6.7b. Almost 50 hours of stability alongside a substantial CO$_2$ yield of 72%, toluene conversion of 95% and a photo-thermal conversion efficacy of 14.1% are accomplished when a value of 116 mW/cm^2 is reached by the infrared irradiation intensity.

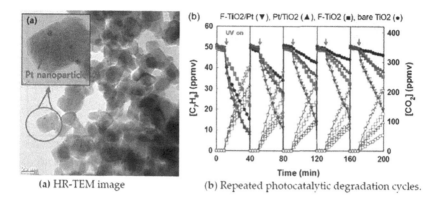

(a) HR-TEM image   (b) Repeated photocatalytic degradation cycles.

FIGURE 6.6   (a) HR-TEM image. (b) Repeated photocatalytic degradation cycles of gaseous toluene on F-TiO$_2$/Pt [46].

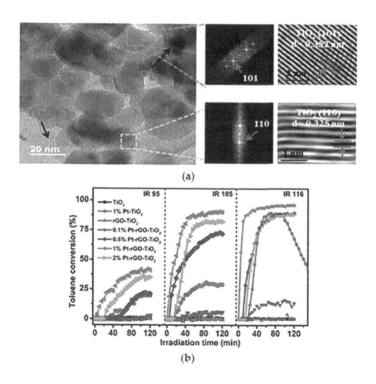

FIGURE 6.7   (a) High-angle annular dark-field scanning transmission electron microscopy images and HRTEM of 1% Pt-rGO-TiO$_2$. (b) Time course of toluene conversion over TiO$_2$, 1% Pt-TiO$_2$ and x% Pt-rGO-TiO$_2$ (x = 0, 0.1, 0.5, 1, and 2) under IR irradiation with various light intensities (95, 106 and 116 cm^2) [47].

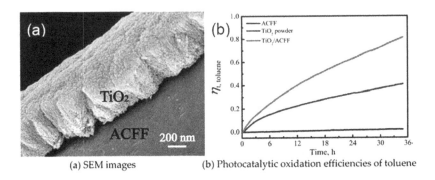

(a) SEM images   (b) Photocatalytic oxidation efficiencies of toluene

FIGURE 6.8   (a) SEM images. (b) Photocatalytic oxidation efficiencies of toluene as function of photocatalytic time under UV irradiation with TiO_2/ACFF porous composites [6].

The photocatalysis performance of metal oxides is improved via their synthesis with adsorption materials. Via the in situ deposition of TiO_2 microspheres within the carbon fibres in ACFF, nanostructured TiO_2/activated carbon fibre-felt (TiO_2/ACFF) porous composites were generated in an academic work [6]. Nanocrystals of TiO_2 microspheres serve to generate hierarchical nanostructures, as displayed in Figure 6.8a. Stellar photodegradation and adsorption characteristics for toluene are present in the TiO_2/ACFF porous composites as a result of the synergetic effects of nanostructured ACFF and TiO_2. Toluene adsorption is accelerated and the TiO_2 band-gap energy is diminished to 2.95 eV via the ACFF present in the TiO_2/ACFF impeding the recombination of electron–hole pairs which, in turn, causes substantially improved photocatalytic properties for toluene.

The examination of TiO_2 and zeolite hybrids for VOC oxidation occurred in a distinct investigation that synthesised decahedral anatase particles (DAPs) with TiO_2 nanoparticles (Ti-NP), reflecting the aforementioned combination [49]. The three forms of TiO_2 comprised 1.0 µm clusters of TiO_2t made of 15 nm average particle size, ca. 100 nm DAPs and 5 nm Ti-NPs. The single titania particles demonstrated roughly 10 times less photoactivity than the composites of zeolitic material with TiO_2-NP added to it. The photocatalytic degradation of aldehydes was resolved by utilising TiO_2-impregnated glass fibre and polyester in a separate study [50]. Additionally, the photocatalytic removal of isovaleric acid and isovaleraldehyde was enabled by the coating of colloidal silica in reactors to the glass fibre tissues fixed to the TiO_2 nanoparticles [51,52].

6.2.2   ZINC OXIDE

Rapid and effective chemical decontamination of VOCs is enacted by zinc oxide, offering an alternative to TiO_2 [53]. Figure 6.9a–d demonstrates the SEM values for the hydrothermal, citrate precursor and solvothermal synthetic techniques used and compared to prepare $ZnAl_2O_4$ for the photocatalytic degradation of gaseous toluene, per another academic work [54]. As portrayed in Figure 6.9, a photocatalytic efficacy

(a–d) SEM images for ZnAl₂O₄ nanoparticles.

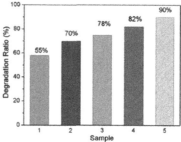
(e) Degradation percentage of toluene.

FIGURE 6.9 SEM images for ZnAl$_2$O$_4$ nanoparticles synthesised with (a) hydrothermal, (b) citrate precursors, and (c, d) solvothermal synthetic methods. (e) The degradation percentage of toluene among 1 (ZnAl$_2$O$_4$ nanoparticles + citrate precursors), 2 (P25 TiO$_2$), 3 (ZnAl$_2$O$_4$ nanoparticles + hydrothermal), 4 (TiO$_2$ nanoballs) and 5 (ZnAl$_2$O$_4$ nanoparticles + solvothermal synthetic) samples under UV illumination [54].

for toluene of roughly 90% was displayed by the ZnAl$_2$O$_4$ samples synthesised via the facile solvothermal technique. A method demonstrating the potential for air purification is the photocatalytic oxidation of gaseous pollutants over UV-illuminated ZnAl$_2$O$_4$.

6.2.3 NICKEL OXIDE

Figure 6.10a–d displays the distinct pyridine volume rations noted within the NiO (NiO/N-doped carbon nanotubes) that reinforce the catalysis performance of NCNTs which were compared in a distinct research paper [55]. As Figure 6.10 portrays, the rise in doped graphitic-like N (NG content of NCNTs) led to a rise in the low-temperature reducibility and oxygen adspecies concentration of NiO/NCNTs. A conclusive conversion of toluene at 248°C possessing a TOF value of 160°C, almost 10 times more than NiO/CNTs, can be accomplished via the maximisation of the NiO-NCNTs catalyst with 6.22% of NG content.

6.2.4 TUNGSTEN TRIOXIDE

VOCs were degraded under visible light via the amalgamation of WO$_3$ functioning as a distinct co-catalyst for Pt and nanodiamond (ND), per another study [56]. Similar to WO$_3$ loaded with Pt and superior to the frequent co-catalysts of WO$_3$, namely CuO, Au, Od and Ag, almost 17 times more photocatalytic activity for the degradation of acetaldehyde was observed in the WO$_3$ loaded with NDs than unmodified WO$_3$. Within the overall photocatalysis process, a crucial part is played by the surface conductivity of the ND incorporated into the WO$_3$. Reduced graphene oxide, graphene oxide and other carbon-based co-catalysts were outperformed regarding photocatalytic activity in comparison to WO$_3$ loaded with NDs.

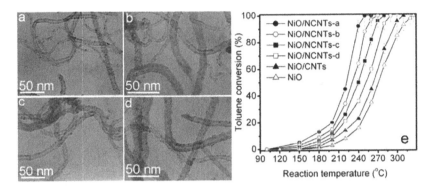

FIGURE 6.10 TEM images of NiO/NCNT catalysts with the pyridine to 3-(aminomethyl) pyridine volume ratios of (a) 5, (b) 3, (c) 1 and (d) 0. (e) Their toluene conversion vs. reaction temperatures against those of NiO/CNTs [55].

6.2.5 MANGANESE OXIDE

Commercial MnO_2 was synthesised with mixed copper manganese oxide (CuO/Mn_2O_3), amorphous manganese oxide (AMO) and cryptomelane-type octahedral molecular sieve (OMS-2) manganese oxide in a certain academic paper [57]. Oxidative activities of commercial MnO_2 were surpassed by CuO/Mn_2O_3, AMO and OMS-2 as a result of redox, morphology, structure and hydrophobicity characteristics among other complex factors. A polyacrylonitrile-based activated carbon nanofiber (PAN-ACNF) support has manganese oxide (MnOx) deposited upon it following the construction of a new hybrid catalyst designed for long-term formaldehyde removal, per another research paper [58]. In conditions void of UV light and either humid or dry, the performance of the PAN-ACNF was two times more effective due to the synergetic effects on the removal of formaldehyde performance provoked by the amalgamation of PAN-ACNF and MnOx. Enhanced catalytic removal of gaseous VOCs was generated by interacting CoAl-mixed oxides and cerium oxide with the manganese oxides [59,60].

6.2.6 BI-BASED COMPOUNDS

For the photocatalytic removal of VOCs, Bi_2WO_6 was amalgamated with highly stable carbon quantum dots (CQDs) in a certain study [61]. Compared with unmodified Bi_2WO_6, the photoexcited charge separation and adsorption into the visible light region are enhanced by the $CQDs/Bi_2WO_6$. In conditions of visible light irradiation and UV-vis, increased photocatalytic oxidation activities towards toluene and acetone were demonstrated by the $CQDs/Bi_2WO_6$ catalyst.

6.2.7 AG-BASED COMPOUNDS

The photocatalytic activity of AgBr was examined in a distinct academic work [62]. While larger crystallites of AG were witnessed, the highest level of H_2 generation activity was observed in the $AgBr(N_2)$ at a temperature of 250°C. Utilising the deposition–precipitation technique, a WO₃ substrate was loaded with AgBr to synthesise a new AgBr/WO₃ composite photocatalyst, in the work by Cao et al. [63]. Under visible light of less than 420 nm, satisfactory levels of photocatalytic activity are demonstrated by AgBr/WO₃.

6.2.8 PLATINUM-SUPPORTED MATERIAL

As portrayed in Figure 6.11a, for use in the catalytic oxidation of BTX, a ceria loading of 10%, 20% and 30% and a Pt loading of 1% were utilised to prepare the Pt/Al₂O₃-CeO₂ according to a separate paper. As presented in Figure 6.11b, roughly 85% of benzene and almost 100% of xylene and toluene were removed by highly active synthesised nanocatalysts, per the findings on toluene oxidation.

Pt particle sizes between 1.2 and 2.2 nm were used within the reduction approach comprising a series of Pt/Al₂O₃ altered with ethylene glycol (EG), per an academic work [65]. At a temperature of 145°C, maximal catalytic oxidation activity of benzene was demonstrated with the Pt/Al₂O₃ catalyst with a Pt size of 1.2 nm. Whether in coexistences with H₂O, cyclohexane, CO₂ or in dry air, throughout the long-term ongoing test, stellar durability and adaptability to distinct VOCs were demonstrated for benzene oxidation by these catalysts. At ambient temperature, the catalytic decomposition of formaldehyde was enacted by adding to an anodic alumite plate the Pt/TiO₂/Al₂O₃ catalyst, per a separate academic work [66]. At such temperatures and within the context of the decomposition of HCHO, satisfactory activity was exhibited by the developed catalyst. In the mentioned conditions, high activity towards the catalytic decomposition of formaldehyde was exhibited by the Pt/TiO₂/Al₂O₃ catalyst, also displaying a rapid activation of absorbing oxygen.

6.2.9 IRIDIUM PARTICLES

The complete oxidation of VOCs was achieved via the use of silica-supporting synthesised iridium particles, causing iridium oxide particle size reductions due to the elevated catalytic activity, per a certain research paper [67].

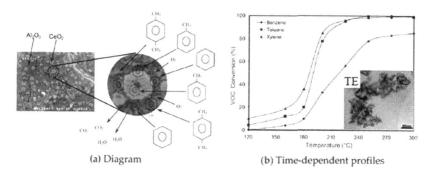

(a) Diagram   (b) Time-dependent profiles

FIGURE 6.11   (a) Reaction mechanism for total oxidation of VOCs over nanostructured Pt/Al₂O₃ CeO₂ catalysts. (b) Comparison of catalytic performance of synthesised Pt (1 wt%)/Al₂O₃-CeO₂ (30 wt%) nanocatalyst for total oxidation of benzene, toluene and xylene [64].

6.3 CARBON-BASED MATERIALS

6.3.1 CARBON-BASED

As demonstrated in Figure 6.12a and b, temperatures between 40 and 150°C were used to examine the performance of a new Pt/carbon nanotube (CNT) catalyst on oxidation of *o*-xylene (BTEX), toluene, benzene and ethylbenzene via a molecular-level mixing technique, in accordance with an academic work [68]. The adsorption ability of multiwalled carbon nanotubes enables increased surface BTEX concentration which, in turn, encourages oxidation activity. The catalyst was unaltered by moisture in the system while the conversion of BTEX at moderately low temperatures with high activity is enabled by the unique hydrophobic feature that the catalyst possesses.

6.3.2 GRAPHENE AND GRAPHENE OXIDE (GO)

High thermal and chemical stability, resilient pore structure, facile synthesis technique, lightweight nature and high surface area are among the advanced properties of graphene oxide (GO) and graphene that lead to its perception as an effective matrix for absorbing gaseous pollutants [69]. Enhancements to the efficacy of VOC removal have been sought after via the incorporation of multiple metal oxides, such as CO_3O_4, WO_3, TiO_2 and SnO_2, into graphene/GO, as reflected by extensive experimenting. In terms of the improvement TiO_2 photocatalytic activity, TiO_2/carbon which can include activated carbon, fullerenes and carbon nanotubes demonstrates equivalent results to the metal oxides such as TiO_2/graphene, despite the latter's unique electronic and structural properties [70]. Given that functionalised carbon nanotubes are costly, more profitability can be found in the use of GO-based hybrid multifunctional materials as low-cost graphite materials and GO can be prepared on a large scale [69].

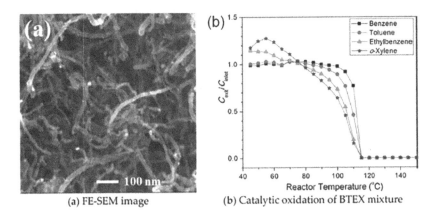

(a) FE-SEM image   (b) Catalytic oxidation of BTEX mixture

FIGURE 6.12 (a) Field emission scanning electron microscopy image. (b) Catalytic oxidation of BTEX mixture as a function of reactor temperature over 30 wt% Pt/CNT catalyst [68].

Within an ethanol–water solvent, a facile hydrothermal reaction of TiO_2 and GO enabled the preparation of the TiO_2/graphene nanocomposites. A reduction in photocatalytic activity occurs due to the greater weight ratio in TiO_2/graphene and unmodified TiO_2 demonstrates significantly less photocatalytic stability and activity towards the gas-phase degradation of benzene than the mentioned nanocomposite. The structure of graphene-based composites can be influenced greatly by photo-catalytic efficacy. The manner in which methanol gas-phase photooxidation was impacted by the co-catalysts few-layer graphene, reduced GO and GO was investigated and compared in an academic work [71]. The degradation of methanol at the highest rate was achieved by the reduced GO.

6.4 APPLICATIONS ON BUILDINGS

Detrimental VOCs are effectively diminished or expunged under UV light irradiation via the significant photoactivated properties of the photocatalyst adopted for use on construction materials. Due to the self-cleaning characteristics of such materials, maintenance costs are reduced while polluting compounds are degraded at the surface of the materials concurrently. Concrete, stone, glass, asphalt, mortars and other such building materials were mixed or coated with these photocatalysts, such as TiO_2, forming the principal function for the application of self-cleaning [72–74].

6.4.1 INDOOR AIR TREATMENT

High quality of air can be sustained via the filtration of VOCs from the air using indoor air treatment. An enhanced performance regarding VOC oxidation is enabled by coating the filter with nanomaterials. The photocatalytic performance can be substantially improved while Pt nanoparticles can be expunged by the ZnO coating in the new Pt. ZnO/SiC filter developed for the oxidation of toluene, per an academic work [75]. Within 10 days at a temperature of 210°C and a filtration velocity of 0.72 m per minute, a total conversion of toluene is accomplished by this filter. Frequently, the material involved in application in this domain is TiO_2. Glass fibre tissue-supported TiO_2 is used to ascertain the efficacy of heterogeneous photocatalysis procedures and non-thermal plasma regarding indoor air treatment, according to a separate research paper [76]. The removal efficacy was found to improve via the amalgamation of photocatalysis and plasma DBD, among the multiple influential elements investigated in the mentioned work. In addition, BTEX removal was found to significantly improve the photocatalytic efficacy of the lanthanide ion doping paired with the TiO_2 catalysts, per the same academic researchers [77]. The highest level of photocatalytic activity was demonstrated by the 1.2% Ln^{3+}–TiO_2 catalysts while the highest level of adsorption capability was demonstrated by the 0.7% Ln^{3+}–TiO_2 catalysts, as revealed by the comparison of multiple lanthanide ion dosages paired with distinct kinds of Ln^{3+}–TiO_2 (La^{3+}/Nd^{3+}–TiO_2). A sparse number of investigations utilised photocatalysts that were not ZnO or TiO_2 which, in turn, were found to be the most frequently utilised materials and catalysts for air purification, in accordance with a 2017 review conducted by another academic [78].

6.4.2 COATING

The coating applied to construction materials consisting of photocatalysts serves multiple functions. This is exemplified by the effective degradation of organic dirt by rainfall and the removal of dust via the enhanced surface hydrophilicity that TiO_2 coating on glass surfaces provides [79]. Pilkington Activ. Clear and Pilkington Activ. Blue comprise two self-cleaning glasses that were analysed regarding their energy efficacy and photocatalytic activity in a certain research paper [80]. In conditions of simulated solar irradiation and ultraviolet irradiation, superior reactivity in the degradation of pre-adsorbed stain and superior activity towards 2-propanol oxidation, respectively, were displayed by the Pilkington Activ. Clear than the Blue version. Between 90% and 98% of decane degradation is attributed to TiO_2-coated exterior paints [81]. Photocatalytic oxidation can be improved by coating materials in air filters with TiO_2 [82]. Carbon cloth fibres coated with TiO_2 (TiO_2/CCFs) and fibre glass fibres coated with TiO_2 (TiO_2/FGFs) within a pilot duct system comprised the two commercially available photocatalytic oxidation air filters whose performances are compared in a separate study [83]. Multiple impacting elements were examined while alcohols, ketones, aromatics and alkanes exhibited the greatest single-pass removal efficacy of such air filters in descending order—alcohol sand alkanes being the most and least effective, respectively. In conditions of visible light, the photocatalytic removal of NO was carried out by coating Fe, F and N ions-doped TiO_2 upon ceramic tiles, per another academic paper [84]. Ceramic tiles displayed a capacity to purify organic compounds and air pollutants of inorganic NO alike. The antibacterial ability that diminishes the chance of bacterial infection, low water adsorption performance, photocatalytic efficacy and good fastness are all improved via the coating placed on the ceramic tiles. Depositing transparent self-cleaning coatings on stone by using spray-coating to apply anatase colloidal suspensions was undertaken in a separate academic work to preserve the aesthetic characteristics of historical architectures and monuments [72]. Untreated travertine was compared with TiO_2 coatings with one and three spray cycles in terms of their photocatalytic performances. Within the context of long-term usage, substantial variances were not discovered between multi-layer and single coatings, meanwhile, the ageing process was found to lead to the degradation of the self-cleaning ability of the mentioned coatings until they displayed low efficacy. An undercoat and novel colloidal TiO_2 with increased stability are required while the long-term preservation of stone appears to benefit from the utilisation of nanostructure TiO_2, per the findings of this paper.

Architectural mortar has a transparent photocatalytic coating comprising TiO_2 particles placed upon it in order to accomplish a resilient weathering-resistant capability and high photocatalytic efficacy, in the work by Guo et al. [85]. During a simulated façade-weathering procedure, the capabilities of the material demonstrated no clear deterioration, and under conditions of visible light irradiation and UV-A, the photocatalytic performances regarding self-cleaning and air-purifying characteristics were demonstrably worse in the TiO_2-intermixed specimens.

6.4.3 PAINTS

Paints are frequently treated with commercial TiO_2, P25 (25% rutile and 75% anatase phases). The surface of self-compacting architectural mortars (SCAM) is directly treated with TiO_2-containing transparent paint in a certain research paper [86]. In all conditions, a resilient weathering resistance capability and a high photocatalytic capacity to remove rhodamine b were demonstrated by the SCAM sample coated with paint containing TiO_2, per the findings of this work. Acid Orange 7 functioned as a model compound to analyse the photocatalytic characteristics of self-cleaning acrylic paint with ZnO and TiO_2 within them, in a separate study [87]. In line with the weathering time, the photocatalytic activity of TiO_2 augments. Nonetheless, the photocorrosion and/or loss of ZnO particles in the weathering process leads to a reduction in photocatalytic activity following weathering, despite the initial substantial rise in the photoactivity of non-weathered paints with ZnO. Mesoporous TiO_2 ($MTiO_2$) is a key component within a synthesised self-cleaning acrylic paint demonstrating durability and efficacy in another academic work [88]. Serious degradation of the binder is seemingly averted via the self-cleaning properties that $MTiO_2$ microspheres deliver to the acrylic paint films, resulting in increased durability and photoactivity in such paints when compared to the reference paint containing P25 incorporated in the cyclic analysis.

Particular applications for indoor or outdoor environments can be specifically matched by self-cleaning photocatalytic paints designed for the context. TiO_2 efficacy within indoor environments confronts difficulty due to the need for UV photons for electron–holes production as a result of its large band gap. Even so, TiO_2 serves as the basis for the majority of photocatalytic paints. In conditions of real indoor light, indoor commercial self-cleaning paints are compared via photocatalytic activity assessments in a distinct academic paper [89]. TiO_2 with distinct crystallographic forms and quantities reside in such paints. The sensitising effect leads to the selective bleaching of pollutant probes and visible light conditions provoke low levels of activity in all samples, per the findings analysing distinct lighting. Resultantly, the removal of the pollutants is significantly impacted by their capacity to inject electrons in the conduction band of TiO_2. Increasing durability via the incorporation into commercial TiO_2 nanopowders in paints, mortars and coatings of sulphuric and nitric acids was analysed as a pre-treatment in another academic work [90]. Sulphuric and nitric acids were compared in terms of their reflectance and photocatalytic performances. Maintaining the long-term photocatalytic activity of TiO_2 nanoparticles while enhancing their wavelength durability and reflectance is demonstrated by the sulphuric acid treatment. On the other hand, a roughly 20% reduction in photocatalytic activity and crystallinity was provoked by nitric acid treatment.

6.4.4 CONSTRUCTION MATERIALS

In the production of coloured mortars, the potential interactions with iron oxide pigments are examined in the analyses of the self-cleaning and photocatalytic activity of coloured mortars containing TiO_2 in an academic work [91]. Inferior soiling in atmospheric exposure is exhibited by mortars containing TiO_2. Nonetheless, when compared to white mortars, lower photocatalytic activity was provoked by the iron oxide pigments. The photocatalyst TiO_2 is applied in the majority of cases with concrete, per a review in this area by another researcher [92]. Future applications of concrete were noted in the form of the relatively novel approach of utilising the semiconductor oxide $LiNbO_3$ which in other cases substitutes TiO_2 in electronic devices for artificial photosynthesis and demonstrates potential. The non-toxic characteristics, high chemical stability and low cost of TiO_2 lead to the sparse usage of other photocatalysts and its predominant usage in applications on buildings as well as its widespread academic review. In accordance with the properties displayed by distinct materials, such materials will be used in diverse future applications.

6.5   SUMMARY

The air-cleaning characteristics of photocatalytic and thermal catalytic materials demonstrate potential in their implementation on construction materials and buildings, reflecting their increasing usage as an effective means of removing VOCs. The non-toxic properties, high chemical stability and low cost make TiO_2 the most popular, effective and economical photocatalyst. VOC removal via the implementation of photocatalysts has predominantly involved the usage of TiO_2 which has been studied extensively over the last few decades. Future applications on buildings are anticipated to use distinct thermal catalysts instead of TiO_2 due to ongoing investigations and the stellar performances displayed by such alternative catalysts. Further amalgamations of distinct nanomaterials to produce additional catalysts are also anticipated given that the synergy with hybrid adsorption materials improves the photocatalysis performance of metal oxides [6]. Within the context of removing VOCs, the technique of intermixing is surpassed by the more widespread and effective technique of coating in the multiple implementations of catalytic nanomaterials on buildings [85]. Future research would do well to focus on enhancing the efficacy of additional applications in particular conditions, such as visible light, to improve the efficacy of removing indoor VOCs.

REFERENCES

1. Spengler, J.D. and Q. Chen. Indoor air quality factors in designing a healthy building. *Annual Review of Energy and the Environment*, 2000, **25**: pp. 567–600.
2. Jones, A.P. Indoor air quality and health. *Atmospheric Environment*, 1999, **33**: pp. 4535–4564.
3. Maroni, M., B. Seifert, and T. Lindvall. *Indoor Air Quality: A Comprehensive Reference Book*, Vol. 3, 1995. New York, NY, USA: Elsevier.
4. Weschler, C.J. Ozone's impact on public health: Contributions from indoor exposures to ozone and products of ozone-initiated chemistry. *Environmental Health Perspect*, 2006, **114**: pp. 1489–1496.
5. Yang, C., G. Miao, Y. Pi, Q. Xia, J. Wu, Z. Li, and J. Xiao. Abatement of various types of VOCs by adsorption/catalytic oxidation: A review. *Chemical Engineering Journal*, 2019, **370**: pp. 1128–1153.
6. Li, M., B. Lu, Q.-F. Ke,, Y.-J. Guo, and Y.-P. Guo. Synergetic effect between adsorption and photodegradation on nanostructured TiO_2/activated carbon fiber felt porous composites for toluene removal. *Journal of Hazardous Materials*, 2017, **333**: pp. 88–98.

7. González-García, P. Activated carbon from lignocellulosics precursors: A review of the synthesis methods characterization techniques and applications. *Renewable and Sustainable Energy Reviews*, 2018, **82**: pp. 1393–1414.

8. Spokas, K.A., J.M. Novak, C.E. Stewart, K.B. Cantrell, M. Uchimiya, M.G. DuSaire, and K.S. Ro. Qualitative analysis of volatile organic compounds on biochar. *Chemosphere*, 2011, **85**: pp. 869–882.

9. Zhang, X., B. Gao, A.E. Creamer, C. Cao, and Y. Li. Adsorption of VOCs onto engineered carbon materials: A review. *Journal of Hazardous Materials*, 2017, **338**: pp. 102–123.

10. Zhang, Z., J. Chen, Y. Gao, Z. Ao, G. Li, T. An, Y. Hu, and Y. Li. A coupled technique to eliminate overall nonpolar and polar volatile organic compounds from paint production industry. *Journal of Cleaner Production*, 2018, **185**: pp. 266–274.

11. Tompkins, D.T. and M.A. Anderson. *Evaluation of Photocatalytic Air Cleaning Capability: A Literature Review and Engineering Analysis*, 2001. Atlanta, GA: ASHRAE.

12. Mo, J., Y. Zhang, Q. Xu, J.J. Lamson, and R. Zhao. Photocatalytic purification of volatile organic compounds in indoor air: A literature review. *Atmospheric Environment*, 2009, **43**: pp. 2229–2246.

13. Shayegan, Z., C.-S. Lee, and F. Haghighat. T_iO_2 photocatalyst for removal of volatile organic compounds in gas phase—A review. *Chemical Engineering Journal*, 2018, **334**: pp. 2408–2439.

14. Cheng, M. and S. Brown. VOCs identified in Australian indoor air and product emission environments. In: *Proceedings of the National Clean Air Conference*, Beijing, China, 4–9 September 2005: pp. 23–27.

15. Wang, S., H.M. Ang, and M.O. Tade. Volatile organic compounds in indoor environment and photocatalytic oxidation: State of the art. *Environment International*, 2007, **33**: pp. 694–705.

16. Kang, D.H., D.H. Choi, S.M. Lee, M.S. Yeo, and K.W. Kim. E_ect of bake-out on reducing VOC emissions and concentrations in a residential housing unit with a radiant floor heating system. *Building and Environment*, 2010, **45**: pp. 1816–1825.

17. Kim, S., Y.-K. Choi, K.-W. Park, and J.T. Kim. Test methods and reduction of organic pollutant compound emissions from wood-based building and furniture materials. *Bioresource Technology*, 2010, **101**: pp. 6562–6568.

18. Zheng, L. and K.W. Shah. Chapter 16: Electrochromic smart windows for green building applications. In: *RSC Smart Materials*, 2019. London: Royal Society of Chemistry: pp. 494–520.

19. Huseien, G.F., A.R.M. Sam, K.W. Shah, M.A. Asaad, M.M. Tahir, and J. Mirza. Properties of ceramic tile waste based alkali-activated mortars incorporating GBFS and fly ash. *Construction and Building Materials*, 2019, **214**: pp. 355–368.

20. Seh, Z.W., S. Liu, S.Y. Zhang, K.W. Shah, and M.Y. Han. Synthesis and multiple reuse of eccentric $Au@TiO_2$ nanostructures as catalysts. *Chemical Communications*, 2011, **47**: pp. 6689–6691.

21. Hussain, M., P. Akhter, J. Iqbal, Z. Ali, W. Yang, N. Shehzad, K. Majeed, R. Sheikh, U. Amjad, and N. Russo. VOCs photocatalytic abatement using nanostructured titania-silica catalysts. *Journal of Environmental Chemical Engineering*, 2017, **5**: pp. 3100–3107.

22. Singh, E., M. Meyyappan, and H.S. Nalwa. Flexible graphene-based wearable gas and chemical sensors. *ACS Applied Materials & Interfaces*, 2017, **9**: pp. 34544–34586.

23. Andre, R.S., R.C. Sanfelice, A. Pavinatto, L.H.C. Mattoso, and D.S. Correa. Hybrid nanomaterials designed for volatile organic compounds sensors: A review. *Material & Design*, 2018, **156**: pp. 154–166.

24. Chen, H., C.E. Nanayakkara, and V.H. Grassian. Titanium dioxide photocatalysis in atmospheric chemistry. *Chemical Reviews*, 2012, **112**: pp. 5919–5948.

25. Kwon, S., M. Fan, A.T. Cooper, and H. Yang. Photocatalytic applications of micro- and nano-TiO$_2$ in environmental engineering. *Critical Reviews in Environmental Science and Technology*, 2008, **38**: pp. 197–226.

26. Tsang, C.H.A., K. Li, Y. Zeng, W. Zhao, T. Zhang, Y. Zhan, R. Xie, D.Y.C. Leung, and H. Huang. Titanium oxide based photocatalytic materials development and their role of in the air pollutants degradation: Overview and forecast. *Environment International*, 2019, **125**: pp. 200–228.

27. Hu, M., Z. Yao, and X. Wang. Graphene-based nanomaterials for catalysis. *Industrial and Engineering Chemistry Research*, 2017, **56**: pp. 3477–3502.

28. Pan, Y., X. Yuan, L. Jiang, H. Yu, J. Zhang, H. Wang, R. Guan, and G. Zeng. Recent advances in synthesis, modification and photocatalytic applications of micro/nano-structured zinc indium sulfide. *Chemical Engineering Journal*, 2018, **354**: pp. 407–431.

29. Sharma, R., S. Sharma, S. Dutta, R. Zboril, and M.B. Gawande. Silica-nanosphere-based organic–inorganic hybrid nanomaterials: Synthesis, functionalization and applications in catalysis. *Green Chemistry*, 2015, **17**: pp. 3207–3230.

30. Huang, H., Y. Xu, Q. Feng, and D.Y. Leung. Low temperature catalytic oxidation of volatile organic compounds: A review. *Catalysis Science & Technology*, 2015, **5**: pp. 2649–2669.

31. Pelaez, M., et al. A review on the visible light active titanium dioxide photocatalysts for environmental applications. *Applied Catalysis B: Environmental*, 2012, **125**: pp. 331–349.

32. Truppi, A., F. Petronella, T. Placido, M. Striccoli, A. Agostiano, M. Curri, and R. Comparelli. Visible-light-active TiO$_2$-based hybrid nanocatalysts for environmental applications. *Catalysts*, 2017, **7**: p. 100.

33. Demeestere, K., J. Dewulf, and H. Van Langenhove. Heterogeneous photocatalysis as an advanced oxidation process for the abatement of chlorinated, monocyclic aromatic and sulfurous volatile organic compounds in air: State of the art. *Critical Reviews in Environmental Science and Technology*, 2007, **37**: pp. 489–538.

34. Fujishima, A. and K. Honda. Electrochemical photolysis of water at a semiconductor electrode. *Nature*, 1972, **238**: pp. 37–38.

35. Maira, A.J., K.L. Yeung, C.Y. Lee, P.L. Yue, and C.K. Chan. Size effects in gas-phase photo-oxidation of trichloroethylene using nanometer-sized TiO$_2$ catalysts. *Journal of Catalysis*, 2000, **192**: pp. 185–196.

36. Maira, A.J., J.M. Coronado, V. Augugliaro, K.L. Yeung, J.C. Conesa, and J. Soria. Fourier transform infrared study of the performance of nanostructured TiO$_2$ particles for the photocatalytic oxidation of gaseous toluene. *Journal of Catalysis*, 2001, **202**: pp. 413–420.

37. Lee, H.J., H.O. Seo, D.W. Kim, K.-D. Kim, Y. Luo, D.C. Lim, H. Ju, J.W. Kim, J. Lee, and Y.D. Kim. A high-performing nanostructured TiO$_2$ filter for volatile organic compounds using atomic layer deposition. *Chemical Communications*, 2011, **47**: pp. 5605–5607.

38. Weon, S. and W. Choi. TiO$_2$ nanotubes with open channels as deactivation-resistant photocatalyst for the degradation of volatile organic compounds. *Environmental Science & Technology*, 2016, **50**: pp. 2556–2563.

39. Weon, S., J. Choi, T. Park, and W. Choi. Freestanding doubly open-ended TiO$_2$ nanotubes for ecientphotocatalytic degradation of volatile organic compounds. *Applied Catalysis B: Environmental*, 2017, **205**: pp. 386–392.

40. Di Paola, A., M. Bellardita, and L. Palmisano. Brookite: The least known TiO$_2$ photocatalyst. *Catalysts*, 2013, **3**: pp. 36–73.

41. Monai, M., T. Montini, and P. Fornasiero. Brookite: Nothing new under the sun? *Catalysts*, 2017, **7**: p. 304.

42. Wu, C., Y. Yue, X. Deng, W. Hua, and Z. Gao. Investigation on the synergetic e_ect between anatase and rutile nanoparticles in gas-phase photocatalytic oxidations. *Catalysis Today*, 2004, **93–95**: pp. 863–869.

43. Chen, K., L. Zhu, and K. Yang. Tricrystalline TiO_2 with enhanced photocatalytic activity and durability for removing volatile organic compounds from indoor air. *Journal of Environmental Sciences*, 2015, **32**: pp. 189–195.

44. Zhong, L., J.J. Brancho, S. Batterman, B.M. Bartlett, and C. Godwin. Experimental and modeling study of visible light responsive photocatalytic oxidation (PCO) materials for toluene degradation. *Applied Catalysis B: Environmental*, 2017, **216**: pp. 122–132.

45. Qiu, X., et al. Hybrid $CuxO/TiO_2$ nanocomposites as risk-reduction materials in indoor environments. *ACS Nano*, 2012, **6**: pp. 1609–1618.

46. Weon, S., J. Kim, and W. Choi. Dual-components modified TiO_2 with Pt and fluoride as deactivation-resistant photocatalyst for the degradation of volatile organic compound. *Applied Catalysis B: Environmental*, 2018, **220**: pp. 1–8.

47. Li, J.J., S.C. Cai, E.Q. Yu, B. Weng, X. Chen, J. Chen, H.P. Jia, and Y.J. Xu. E_cient infrared light promoted degradation of volatile organic compounds over photo-thermal responsive Pt-rGO-TiO_2 composites. *Applied Catalysis B: Environmental*, 2018, **233**: pp. 260–271.

48. Jo, W.-K. and J.-T. Kim. Application of visible-light photocatalysis with nitrogen-doped or unmodified titanium dioxide for control of indoor-level volatile organic compounds. *Journal of Hazardous Materials*, 2009, **164**: pp. 360–366.

49. Suárez, S., I. Jansson, B. Ohtani, and B. Sánchez. From titania nanoparticles to decahedral anatase particles: Photocatalytic activity of TiO_2/zeolite hybrids for VOCs oxidation. *Catalysis Today*, 2019, **326**: pp. 2–7.

50. Elfalleh, W., A.A. Assadi, A. Bouzaza, D. Wolbert, J. Kiwi, and S. Rtimi. Innovative and stable TiO_2 supported catalytic surfaces removing aldehydes under UV-light irradiation. *Journal of Photochemistry and Photobiology A: Chemistry*, 2017, **343**: pp. 96–102.

51. Assadi, A.A., A. Bouzaza, Wolbert, D., and P. Petit. Isovaleraldehyde elimination by UV/TiO_2 photocatalysis: Comparative study of the process at different reactors configurations and scales. *Environmental Science and Pollution Research*, 2014, **21**: pp. 11178–11188.

52. Palau, J., A.A. Assadi, J. Penya-Roja, A. Bouzaza, D. Wolbert, and V. Martínez-Soria. Isovaleraldehyde degradation using UV photocatalytic and dielectric barrier discharge reactors, and their combinations. *Journal of Photochemistry and Photobiology A: Chemistry*, 2015, **299**: pp. 110–117.

53. Azzouz, I., Y.G. Habba, M. Capochichi-Gnambodoe, F. Marty, J. Vial, Y. Leprince-Wang, and T. Bourouina. Zinc oxide nano-enabled microfluidic reactor for water purification and its applicability to volatile organic compounds. *Microsystems & Nanoengineering*, 2018, **4**.

54. Li, X., Z. Zhu, Q. Zhao, and L. Wang. Photocatalytic degradation of gaseous toluene over $ZnAl_2O_4$ prepared by different methods: A comparative study. *Journal of Hazardous Materials*, 2011, **186**: pp. 2089–2096.

55. Jiang, S., E.S. Handberg, F. Liu, Y. Liao, H. Wang, Z. Li, and S. Song. E_ect of doping the nitrogen into carbon nanotubes on the activity of NiO catalysts for the oxidation removal of toluene. *Applied Catalysis B: Environmental*, 2014, **160–161**: pp. 716–721.

56. Kim, H.I., H.N. Kim, S. Weon, G.H. Moon, J.H. Kim, and W. Choi. Robust co-catalytic performance of nanodiamonds loaded on WO_3 for the decomposition of volatile organic compounds under visible light. *ACS Catalysis*, 2016, **6**(12): pp. 8350–8360.

57. Genuino, H.C., S. Dharmarathna, E.C. Njagi, M.C. Mei, and S.L. Suib. Gas-phase total oxidation of benzene, toluene, ethylbenzene, and xylenes using shape-selective manganese oxide and copper manganese oxide catalysts. *Journal of Physical Chemistry C*, 2012, **116**: pp. 12066–12078.

58. Miyawaki, J., G.H. Lee, J. Yeh, N. Shiratori, T. Shimohara, I. Mochida, and S.H. Yoon. Development of carbon-supported hybrid catalyst for clean removal of formaldehyde indoors. *Catalysis Today*, 2012, **185**: pp. 278–283.

59. Chen, J., X. Chen, D. Yan, M. Jiang, W. Xu, H. Yu, and H. Jia. A facile strategy of enhancing interaction between cerium and manganese oxides for catalytic removal of gaseous organic contaminants. *Applied Catalysis B: Environmental*, 2019, **250**: pp. 396–407.

60. Zhao, Q., Q. Liu, C. Song, N. Ji, D. Ma, and X. Lu. Enhanced catalytic performance for VOCs oxidation on the CoAlO oxides by $KMnO_4$ doped on facile synthesis. *Chemosphere*, 2019, **218**: pp. 895–906.

61. Qian, X., D. Yue, Z. Tian, M. Reng, Y. Zhu, M. Kan, T. Zhang, and Y. Zhao. Carbon quantum dots decorated Bi_2WO_6 nanocomposite with enhanced photocatalytic oxidation activity for VOCs. *Applied Catalysis B: Environmental*, 2016, **193**: pp. 16–21.

62. Kobayashi, B., R. Yamamoto, H. Ohkita, T. Mizushima, A. Hiraishi, and N. Kakuta. Photocatalytic activity of AgBr as an environmental catalyst. *Topics in Catalysis*, 2013, **56**: pp. 618–622.

63. Cao, J., B. Luo, H. Lin, and S. Chen. Photocatalytic activity of novel $AgBr/WO_3$ composite photocatalyst under visible light irradiation for methyl orange degradation. *Journal of Hazardous Materials*, 2011, **190**: pp. 700–706.

64. Abbasi, Z., M. Haghighi, E. Fatehifar, and S. Saedy. Synthesis and physicochemical characterizations of nanostructured $Pt/Al_2O_3-CeO_2$ catalysts for total oxidation of VOCs. *Journal of Hazardous Materials*, 2011, **186**: pp. 1445–1454.

65. Chen, Z., J. Mao, and R. Zhou. Preparation of size-controlled Pt supported on Al_2O_3 nanocatalysts for deep catalytic oxidation of benzene at lower temperature. *Applied Surface Science*, 2019, **465**: pp. 15–22.

66. Wang, L., M. Sakurai, and H. Kameyama. Study of catalytic decomposition of formaldehyde on Pt/TiO_2 alumite catalyst at ambient temperature. *Journal of Hazardous Materials*, 2009, **167**: pp. 399–405.

67. Schick, L., R. Sanchis, V. González-Alfaro, S. Agouram, J.M. López, L. Torrente-Murciano, T. García, and B. Solsona. Size-activity relationship of iridium particles supported on silica for the total oxidation of volatile organic compounds (VOCs). *Chemical Engineering Journal*, 2019, **366**: pp. 100–111.

68. Joung, H.-J., J.-H. Kim, J.-S. Oh, D.-W. You, H.-O. Park, and K.-W. Jung. Catalytic oxidation of VOCs over CNT-supported platinum nanoparticles. *Applied Surface Science*, 2014, **290**: pp. 267–273.

69. Samaddar, P., Y.-S. Son, D.C.W. Tsang, K.-H. Kim, and S. Kumard. Progress in graphene-based materials as superior media for sensing, sorption, and separation of gaseous pollutants. *Coordination Chemistry Reviews*, 2018, **368**: pp. 93–114.

70. Zhang, Y., Z.-R. Tang, X. Fu, and Y.-J. Xu. TiO_2-graphene nanocomposites for gas-phase photocatalytic degradation of volatile aromatic pollutant: Is TiO_2-graphene truly di_erent from other TiO_2-carbon composite materials? *ACS Nano*, 2010, **4**: pp. 7303–7314.

71. Roso, M., C. Boaretti, M.G. Pelizzo, A. Lauria, M. Modesti, and A. Lorenzetti. Nanostructured photocatalysts based on di_erent oxidized graphenes for VOCs removal. *Industrial and Engineering Chemistry Research*, 2017, **56**: pp. 9980–9992.

72. Munafò, P., E. Quagliarini, Go_redo, G.B., F. Bondioli, and A. Licciulli. Durability of nano-engineered TiO_2 self-cleaning treatments on limestone. *Construction and Building Materials*, 2014, **65**: pp. 218–231.

73. Toro, C., B.T. Jobson, L. Haselbach, S. Shen, and S.H. Chung. Photoactive roadways: Determination of CO, NO and VOC uptake coe_cients and photolabile side product yields on TiO_2 treated asphalt and concrete. *Atmospheric Environment*, 2016, **139**: pp. 37–45.

74. Huseien, G.F., K.W. Shah, and A.R.M. Sam. Sustainability of nanomaterials based self-healing concrete: An all-inclusive insight. *Journal of Building Engineering*, 2019, **23**: pp. 155–171.

75. Li, L., F. Zhang, Z. Zhong, M. Zhu, C. Jiang, J. Hu, and W. Xing. Novel synthesis of a high-performance Pt/ZnO/SiC filter for the oxidation of toluene. *Industrial and Engineering Chemistry Research*, 2017, **56**: pp. 13857–13865.

76. Zadi, T., A.A. Assadi, N. Nasrallah, R. Bouallouche, P.N. Tri, A. Bouzaza, M.M. Azizi, R. Maachi, and D. Wolbert. Treatment of hospital indoor air by a hybrid system of combined plasma with photocatalysis: Case of trichloromethane. *Chemical Engineering Journal*, 2018, **349**: pp. 276–286.

77. Li, F.B., X.Z. Li, C.H. Ao, S.C. Lee, and M.F. Hou. Enhanced photocatalytic degradation of VOCs using Ln^{3+}–TiO_2 catalysts for indoor air purification. *Chemosphere*, 2005, **59**: pp. 787–800.

78. Boyjoo, Y., H. Sun, J. Liu, V.K. Pareek, and S. Wang. A review on photocatalysis for air treatment: From catalyst development to reactor design. *Chemical Engineering Journal*, 2017, **310**: pp. 537–559.

79. Chabas, A., T. Lombardo, H. Cachier, M.H. Pertuisot, K. Oikonomou, R. Falcone, M. Verità, and F. Geotti-Bianchini. Behaviour of self-cleaning glass in urban atmosphere. *Building and Environment*, 2008, **43**: pp. 2124–2131.

80. Oladipo, H., C. Garlisi, K. Al-Ali, E. Azar, and G. Palmisano. Combined photocatalytic properties and energy efficiency via multifunctional glass. *Journal of Environmental Chemical Engineering*, 2019, **7**(2): p. 102980.

81. Monteiro, R.A.R., F.V.S. Lopes, A.M.T. Silva, J. Ângelo, G.V. Silva, A.M. Mendes, R.A.R. Boaventura, and V.J.P. Vilar. Are TiO_2-based exterior paints useful catalysts for gas-phase photooxidation processes? A case study on n-decane abatement for air detoxification. *Applied Catalysis B: Environmental*, 2014, **147**: pp. 988–999.

82. Martinez, T., A. Bertron, G. Escadeillas, E. Ringot, and V. Simon. BTEX abatement by photocatalytic TiO_2-bearing coatings applied to cement mortars. *Building and Environment*, 2014, **71**: pp. 186–192.

83. Zhong, L., F. Haghighat, C.S. Lee, and N. Lakdawala. Performance of ultraviolet photocatalytic oxidation for indoor air applications: Systematic experimental evaluation. *Journal of Hazardous Materials*, 2013, **261**: pp. 130–138.

84. Jiang, Q., T. Qi, T. Yang, and Y. Liu. Ceramic tiles for photocatalytic removal of NO in indoor and outdoor air under visible light. *Building and Environment*, 2019, **158**: pp. 94–103.

85. Guo, M.-Z., A. Maury-Ramirez, and C.S. Poon. Photocatalytic activities of titanium dioxide incorporated architectural mortars: Effects of weathering and activation light. *Building and Environment*, 2015, **94**: pp. 395–402.

86. Guo, M.-Z., A. Maury-Ramirez, and C.S. Poon. Self-cleaning ability of titanium dioxide clear paint coated architectural mortar and its potential in field application. *Journal of Cleaner Production*, 2016, **112**: pp. 3583–3588.

87. Baudys, M., J. Krýsa, M. Zlámal, and A. Mills. Weathering tests of photocatalytic facade paints containing ZnO and TiO_2. *Chemical Engineering Journal*, 2015, **261**: pp. 83–87.

88. Amorim, S.M., J. Suave, L. Andrade, A.M. Mendes, H.J. José, and R.F.P.M. Moreira. Towards an efficient and durable self-cleaning acrylic paint containing mesoporous TiO_2 microspheres. *Progress in Organic Coatings*, 2018, **118**: pp. 48–56.

89. Galenda, A., F. Visentin, R. Gerbasi, M. Favaro, A. Bernardi, and N. El Habra. Evaluation of self-cleaning photocatalytic paints: Are they effective under actual indoor lighting systems? *Applied Catalysis B: Environmental*, 2018, **232**: pp. 194–204.

90. Paolini, R., D. Borroni, M. Pedeferri, and M.V. Diamanti. Self-cleaning building materials: The multifaceted effects of titanium dioxide. *Construction and Building Materials*, 2018, **182**: pp. 126–133.

91. Diamanti, M.V., B. Del Curto, M. Ormellese, and M. Pedeferri. Photocatalytic and self-cleaning activity of colored mortars containing TiO_2. *Construction and Building Materials*, 2013, **46**: pp. 167–174.

92. Nath, R.K., M.F.M. Zain, and M. Jamil. An environment-friendly solution for indoor air purification by using renewable photocatalysts in concrete: A review. *Renewable and Sustainable Energy Reviews*, 2016, **62**: pp. 1184–1194.

Index

acid 7, 35, 74–79, 81, 139
alkali-activated 35, 47–50, 52
antimicrobial 26, 130

binder 27–29, 41, 45, 47–50, 111, 139
biological 2, 125
building 1, 25, 35, 67, 84, 103, 114, 140
built cities 15, 103

calcium 28, 30, 41, 48, 70, 96
capsules 67, 69, 73
carbon dioxide 23, 84, 125
carbon nanotube 6, 26, 68, 87, 91, 98, 133
cement 2, 23, 27, 34, 38, 41, 44, 52
clinker 23, 44
coating 4, 7, 11, 25, 103, 107, 110, 138
composites 14, 27, 41, 44, 71
compressive strength 30, 49, 50, 93
concrete 15, 25, 30, 37, 41, 45
construction 12, 15, 24, 35, 42, 49, 93, 104
core-shell 1–4, 7, 9, 14, 16, 120
cost 3, 24, 42, 71, 76, 80, 108
crack 32, 42, 52, 61

deterioration 28, 36, 97, 122, 138
development 1, 6, 16, 23, 42, 48, 68, 84
ductility 40, 46
durability 10, 15, 25, 35, 45, 50, 129

eco-friendly 35, 52, 114
economic 17, 23, 35, 90, 140
economically 17
elevated temperatures 62
energy 4, 12, 23, 35, 42, 67, 74, 83, 120
environment 4, 17, 38, 90, 95, 105, 125

flexural strength 32, 42, 49, 93
fresh properties 43

gels 48, 60
geopolymer 44, 46, 48, 50
glass 9, 25, 42, 91, 95, 103–109, 116–119
graphene 7, 37, 42, 68, 70, 78, 80
greenhouse 23, 44, 111
gypsum 24, 34

hydration 28, 31, 34, 39, 46, 50

indoor 67, 76, 103, 105, 112, 125, 129, 137
infrastructure 15, 23, 41
inorganic 2, 10, 24, 49, 67, 97, 105, 138

mechanical properties 33, 44, 46, 50, 95
metal-coated 69
micro 2, 5, 10
microstructure 25, 27, 41, 46, 48, 51, 80
mineral 55, 60, 100
modulus 41, 92
mortar 21, 29, 31, 34, 38

nano additive 34
nanocomposite 45, 59, 70, 74, 87, 99, 114
nanofiber 55, 68, 83, 134
nanomaterials 16, 24, 30, 38, 83, 110, 115
nanoparticles 2, 9, 24, 29, 34, 46, 73, 87
nanoplate 42, 45, 68, 80, 113
nanopolymer 87, 91, 93, 95, 98
nanoscience 24, 35
nanosensors 97, 99, 105
nanosilica 27, 101
nanosized 49, 67, 75, 77, 83
nanotitanium 45, 63
nanotubes 6, 42, 55, 68, 82
natural 20, 44, 94, 97, 105, 115

organic 6, 10, 24, 49, 50, 67, 78, 97

paraffin 67, 72, 75, 78, 80, 83
phase change materials 67–69, 81, 83–84
pigment 1–2, 7, 9, 12, 15, 17
purification 101, 125, 133, 137, 141

reflectivity 13, 109, 112, 116

silicon 5, 24, 44
solar energy 42, 84, 95, 97, 99, 106, 110
sustainability 24, 26, 33, 52

thermal conductivity 67, 69, 71, 74, 76, 80
titanium 5, 9, 49, 50, 119
tungsten oxide 113, 115, 120

workability 25, 29, 52, 61

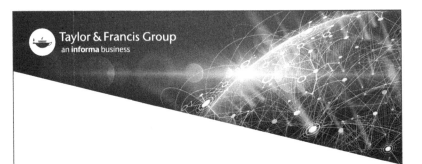

Printed in the United States
by Baker & Taylor Publisher Services